Directed Reading A

Section: Electric Charge and Static Electricity

ELECTRIC CHARGE

_______ **1.** What do you call the tiny particles that make up matter?
  a. electricity
  b. atoms
  c. electrons
  d. charges

_______ **2.** Atoms are made up of protons, neutrons, and what third particle?
  a. charges
  b. electricity
  c. electrons
  d. forces

3. What three types of charge can an object have?

____

____

4. What does the law of electric charges state?

____

____

5. Protons are _______________________ charged.

6. Electrons are _______________________ charged.

7. The force between charged objects is a(n) _______________________.

8. What two things affect the size of the electric force?

____

____

9. The region around a charged object where an electric force is exerted on

another charged object is the _______________________.

10. How do charged objects within an electric field interact?

____

____

CHARGE IT!

11. Why are atoms uncharged?

____

____

12. What happens when an object loses electrons?

____

13. What happens when an object gains electrons?

____

Match the correct definition with the correct term. Write the letter in the space provided.

_______**14.** This happens when electrons are "wiped" from one object to another.

_______**15.** This happens when charges in an uncharged metal object are rearranged without direct contact with a charged object.

_______**16.** This happens when electrons move by direct contact.

a. conduction

b. induction

c. friction

17. When you charge objects by any method, no charges are

_______________________ or _______________________.

18. You can use a device called a(n) _______________________ to see if something is charged.

19. Can you tell if an object has a positive or negative charge with an electroscope?

____

MOVING CHARGES

_______**20.** Which of the following is a material in which charges can move easily?
    a. electrical conductor   **c.** electrical jumper
    b. electrical insulator   **d.** electrical stopper

_______**21.** Which of the following is a material in which charges CANNOT move easily?
    a. electrical conductor   **c.** electrical jumper
    b. electrical insulator   **d.** electrical stopper

_______**22.** Which of the following is a good conductor?
    a. wood   **c.** copper
    b. air   **d.** glass

_______**23.** Which of the following is a good insulator?
    a. aluminum   **c.** mercury
    b. glass   **d.** copper

Directed Reading A *continued*

24. Why are most metals good conductors?

____

____

25. What factors make a material a good insulator?

____

____

STATIC ELECTRICITY

26. What is static electricity?

____

____

27. The loss of static electricity as charges move off an object is called

_______________________.

28. What three things might you notice after an electric discharge?

____

____

____

29. How does lightning occur within a cloud?

____

____

____

30. Why is it unsafe to be at the beach during a lightning storm?

____

____

31. How do lightning rods protect buildings from lightning?

____

____

_________________________ manly continued

Skills Worksheet

Directed Reading A

Section: Electric Current and Electrical Energy

1. The energy of electric charges is called _________________________.

ELECTRIC CURRENT

_______ 2. What is the rate at which charges pass a given point?

    a. electrical energy   **c.** electric current

    b. amps   **d.** voltage

_______ 3. Which of the following is the unit for electric current?

    a. amperes, or amps   **c.** volts

    b. ohms   **d.** charges

_______ 4. Which of these letters is used to represent current in equations?

    a. C **c.** R

    b. I **d.** T

_______ 5. What are the two types of electric current?

    a. electrons and neutrons   **c.** current and electric

    b. alternating and direct   **d.** charge and negative

6. When you flip the switch on a flashlight, a(n) _________________________ is set up in the wire.

7. When charges continually shift from flowing in one direction to flowing in the reverse direction, there is a(n) _________________________ current.

8. When charges always flow in the same direction, there is a(n) _________________________ current.

9. Which type of current is used in batteries and which is used in household outlets?

____

____

____

____

❙ Directed Reading A *continued*

VOLTAGE

_______**10.** What is the potential difference between two points in a circuit called?
  a. resistance   **c.** voltage
  b. current   **d.** charge

_______**11.** Which letter is used to represent voltage in equations?
  a. G **c.** V
  b. C **d.** I

_______**12.** What happens to electric current if voltage becomes larger?
  a. Current decreases.   **c.** Current stays the same.
  b. Current increases.   **d.** No current will flow.

RESISTANCE

_______**13.** What is the opposition to the flow of electric charge called?
  a. resistance   **c.** voltage
  b. current   **d.** charge

_______**14.** Which letter is used to represent resistance in equations?
  a. E **c.** V
  b. C **d.** R

15. Resistance is expressed in _____________________.

16. The higher the resistance of a material is, the lower the

_____________________.

17. What four things determine an object's resistance?

____

____

____

18. Good conductors have a(n) _______________ resistance.

19. Why are high resistance materials useful in light bulbs?

____

____

20. What is a superconductor?

____

____

| Directed Reading A *continued*

GENERATING ELECTRICAL ENERGY

Match the correct definition with the correct term. Write the letter in the space provided.

_______**21.** This changes chemical or radiant energy into electrical energy.

_______**22.** This is a mixture of chemicals in a cell.

_______**23.** This converts light energy into electrical energy.

_______**24.** This conducts thermal energy into electrical energy.

_______**25.** This is the part of the cell through which charges enter and exit.

a. cell

b. electrode

c. thermocouple

d. electrolyte

e. photocell

26. Chemical changes between the electrolyte and the electrodes convert

_____________________ into _____________________ .

27. Compare electrolytes found in a wet cell with those found in a dry cell.

____

____

28. Thermocouples are useful for monitoring the temperatures of what three things?

____

____

____

29. How do photocells convert light energy into electrical energy?

____

____

____

____

Name _______________________________ Class _______________ Date _______________

Directed Reading A

Section: Electrical Calculations

CONNECTING CURRENT, VOLTAGE, AND RESISTANCE

_______ **1.** What is the ratio of voltage to current?
    a. electrical power   **c.** resistance
    b. electrical energy   **d.** current

_______ **2.** Which of the following equations is Ohm's law?
    a. $V = I \times R$ **c.** $R = I \times V$
    b. $I = V \times R$ **d.** $R = V \times I$

3. How did Georg Ohm study the resistances of different materials?

____

____

____

ELECTRIC POWER

_______ **4.** Which of the following is the rate at which electrical energy is changed
into other forms of energy?
    a. electric current   **c.** voltage
    b. electric power   **d.** kilowatt

_______ **5.** In the formula $P = V \times I$, what does the P stand for?
    a. performance   **c.** price
    b. power   **d.** penny

_______ **6.** The rate at which an electrical device uses power is called its
    a. energy rating.
    b. power rating.
    c. star rating.
    d. electrical rating.

_______ **7.** How can you locate the power rating of an electrical device?
    a. Call the electric company.
    b. Look at the label.
    c. Contact the EPA.
    d. Check the voltage.

8. Name two common units of power.

____

____

9. What happens to a light bulb as power increases?

____

____

____

10. One kilowatt is equal to _____________________ W.

MEASURING ELECTRICAL ENERGY

11. What two factors does the electric company use to determine how much a business will pay for electrical energy?

____

____

12. What is the formula for finding electrical energy?

____

13. What unit is usually used to express electrical energy?

____

14. Electric companies use a(n) _____________________ to determine how many kilowatt-hours of energy are used by a household.

15. Name two ways you can help to save energy.

____

____

16. How much of a home's total energy consumption could be made up by lighting alone?

____

17. Why are fluorescent light bulbs more efficient than incandescent light bulbs?

____

____

18. What does the ENERGY STAR® program promote?

____

____

Directed Reading A

Section: Electric Circuits

1. A closed pathway where the start and end point are the same is called

a(n) ____________________.

PARTS OF AN ELECTRIC CIRCUIT

______ **2.** Which of the following is a complete, closed path through which
electric charges flow?
a. electric current
b. electric circuit
c. energy source
d. load

______ **3.** Which of the following is NOT a basic part of a circuit?
a. wires
b. force
c. energy source
d. load

______ **4.** Which of the following is connected to the energy source by wires and
changes electrical energy into other forms of energy?
a. wires
b. force
c. energy source
d. load

5. Name four possible energy sources for a circuit.

____

____

____

____

6. A circuit is opened and closed using a(n) ____________________.

7. What will happen to the charges in a circuit when a switch is closed?

____

____

____

8. What will happen to the charges in a circuit when a switch is open?

____

____

____

Directed Reading A *continued*

TYPES OF CIRCUITS

9. All of the electrical devices in your home are _____________________ in a large circuit.

10. A circuit in which all parts are connected in a single loop is called

a(n) _____________________.

11. What happens to a series circuit if you add more loads?

____

____

____

12. In a series circuit, what happens if there is a break in the circuit?

____

____

13. Why aren't series circuits a convenient way to wire a home?

____

____

____

14. A circuit in which loads are connected side by side is called a(n)

_____________________.

15. Each load in a parallel circuit uses the same _____________________.

16. What happens in a parallel circuit if one load is broken or missing?

____

____

____

17. Why can you use one light or appliance at a time in a parallel circuit even if a load fails?

____

____

____

____

| Directed Reading A *continued*

HOUSEHOLD CIRCUIT SAFETY

______**18.** What is the standard voltage per branch in a home in the United States?
  a. 100 V   **c.** 120 V
  b. 110 V   **d.** 130 V

19. Name two things that can cause a short circuit in your home.

____

____

20. What happens to a fuse to stop the flow of charges through it?

____

____

____

21. A switch that automatically opens if the current is too high is

a(n)________________________.

22. How does a ground fault circuit interrupter act like a small circuit breaker?

____

____

____

23. Name two safety measures to follow when using electrical energy.

____

____

Directed Reading B

Section: Electric Charge and Static Electricity
ELECTRIC CHARGE

Circle the letter of the best answer for each question.

1. What do you call the tiny particles that make up matter?

    a. electricity   **c.** electrons

    b. atoms   **d.** charges

2. Atoms are made of three things, protons, neutrons and what other particle?

    a. charges   **c.** electrons

    b. electricity   **d.** forces

Charges Exert Forces

3. Which of the following laws says that opposite charges attract?

    a. the law of electrons

    b. the law of electric charges

    c. the law of atomic charges

    d. the law of charged atoms

4. Which of the following affects the size of an electric force?

    a. number of atoms

    b. the type of charge

    c. the amount of each charge

    d. the number of protons

The Force Between Protons and Electrons

5. What type of charge do protons have?

    a. negative charge

    b. opposite charge

    c. positive charge

    d. no charge

Circle the letter of the best answer for each question.

6. What type of charge do electrons have?

   a. opposite charge

   b. negative charge

   c. positive charge

   d. no charge

The Electric Force and the Electric Field

7. What is an electric force?

   a. the force between charged objects

   b. the law about charges

   c. the distance between the charges

   d. the amount of each charge

8. What is an electric field?

   a. the area around an atom

   b. the area around an electron

   c. the area around a proton

   d. the area around a charged object

CHARGE IT!

9. When an object loses electrons, what happens to its charge?

   a. The charge becomes positive.

   b. The charge becomes negative.

   c. The size of the charge gets bigger.

   d. The size of the charge gets smaller.

10. When an object gains electrons, what happens to its charge?

   a. The charge becomes positive.

   b. The charge becomes negative.

   c. The size of the charge gets bigger.

   d. The size of the charge gets smaller.

Directed Reading B *continued*

Read the description. Then, <u>draw a line</u> from the dot next to each description to the matching word.

11. charging by wiping electrons from one object to another ●

12. used to test if something is charged ●

13. rearranging charges in an uncharged metal object without direct contact with a charged object ●

14. charging when electrons move by direct contact ●

a. conduction

b. induction

c. friction

d. electroscope

MOVING CHARGES
Conductors
<u>Circle the letter</u> of the best answer for each question.

15. Which of the following is a material in which charges can move easily?

   a. electrical conductor

   b. electrical insulator

   c. electrical maker

   d. electrical machine

Insulators

16. Which of the following is a material in which charges cannot move easily?

   a. electrical conductor

   b. electrical insulator

   c. electrical maker

   d. electrical machine

| Directed Reading B *continued*

STATIC ELECTRICITY

<u>Circle the letter</u> of the best answer for each question.

17. Which of the following is electric charge at rest?

    a. electric current

    b. electric discharge

    c. static electricity

    d. conduct charge

Electric Discharge

18. What do you call the loss of static electricity as charges move off an object?

    a. electric current

    b. electric discharge

    c. static electricity

    d. conduct charge

Lightning Dangers

Read the words in the box. Read the sentences. <u>Fill in each blank</u> with the word or phrase that best completes the sentence.

lightning rod	lightning	grounded

19. Electric discharge within a cloud causes

_______________________.

20. When a building is hit by lightning, a

_______________________ can protect it from damage.

21. The wires in a lightning rod are _______________________ so

The charges go to Earth.

Directed Reading B

Section: Electric Current and Electrical Energy

Circle the letter of the best answer for each question.

1. What is the flow of electric charges called?

   a. electric charge

   b. electric switch

   c. electricity

   d. electric current

ELECTRIC CURRENT

2. What is the rate at which charges pass a given point?

   a. electrical energy

   b. amps

   c. electric current

   d. volts

3. What is the unit used to express electric current called?

   a. amps

   b. ohms

   c. volts

   d. charges

4. Which of these letters is the symbol for current in equations?

   a. C

   b. I

   c. R

   d. T

| Directed Reading B *continued*

Making Changes Move

Read the words in the box. Read the sentences. <u>Fill in each blank</u> with the word that best completes the sentence.

switch	alternating	direct	electron

5. A single _________________________ may take one hour to

travel 1 m through wire.

6. Charges shift from flowing in one direction to another in a(n)

_________________________ current.

7. When you flip a(n) _________________________ on a flashlight,

an electric field is set up.

8. Charges always flow in the same direction in a(n)

_________________________ current.

VOLTAGE

<u>Circle the letter</u> of the best answer for each question.

9. What is the potential difference between two points in a circuit called?

   a. voltage   **c.** amps

   b. cells   **d.** current

10. Which letter is used to represent voltage in equations?

   a. G **c.** V

   b. C **d.** I

Voltage and Electric Current

11. What happens to electric current as voltage gets larger?

   a. Current gets smaller.   **c.** Current drops to 0 A.

   b. Current gets larger.   **d.** Current does not change.

Directed Reading B *continued*

Varying Nature of Voltage

Circle the letter of the best answer for each question.

12. How many volts are televisions, toasters, and alarm clocks made to run on?

a. 600 V

b. 1 V

c. 0 V

d. 120 V

RESISTANCE

13. What is the opposition to the flow of an electric charge called?

a. resistance

b. lightning

c. amps

d. static electricity

14. What unit is used to express resistance?

a. amps

b. volts

c. ohms

d. yards

15. Which of these letters is the symbol for resistance in equations?

a. C

b. R

c. T

d. O

Resistance and Temperature

16. What do you call a material that has a resistance of 0 ohms?

a. conductor

b. resistor

c. insulator

d. superconductor

Directed Reading B *continued*

GENERATING ELECTRICAL ENERGY

Circle the letter of the best answer for each question.

17. What do cells change chemical or radiant energy into?

a. mechanical energy

b. kinetic energy

c. electrical energy

d. thermal energy

Parts of a Cell

18. What change takes place between the electrolyte and the electrodes?

a. chemical energy into electrical energy

b. electrical energy into chemical energy

c. thermal energy into electrical energy

d. electrical energy into thermal energy

Kinds of Cells

Read the description. Then, draw a line from the dot next to each description to the matching word.

19. a mixture of chemicals in a cell ●

20. this converts chemical energy ● into electrical energy

21. the part of the cell through ● which charges enter and exit

a. cell

b. electrolyte

c. electrode

---

22. a cell that has a solid electrolyte ●

23. this converts thermal energy ● into electrical energy

24. this converts light energy into ● electrical energy

25. a cell that has a liquid ● electrolyte

a. photocell

b. dry cell

c. wet cell

d. thermocouple

Directed Reading B

Section: Electrical Calculations
CONNECTING CURRENT, VOLTAGE, AND RESISTANCE
Ohm's Law

Circle the letter of the best answer for each question.

1. Which of the following equations is Ohm's law?

   a. $V = I \times R$

   b. $I = V \times R$

   c. $R = I \times V$

   d. $R = V \times I$

2. As resistance goes up what goes down?

   a. voltage

   b. charge

   c. current

   d. temperature

3. What happens to current as voltage goes up?

   a. Current goes up.

   b. Current stays the same.

   c. Current goes down.

   d. Current drops to 0 A.

ELECTRIC POWER

4. What is the rate at which electrical energy is changed into other forms of energy?

   a. voltage

   b. electrical energy

   c. electric power

   d. amps

| Directed Reading B *continued*

Circle the letter of the best answer for each question.

5. In the formula $P = V \times I$, what does the P stand for?

   a. performance

   b. power

   c. price

   d. penny

Watt: The Unit of Power

6. Which of the following is a unit of power?

   a. kilowatt

   b. volt

   c. joule

   d. amp

7. One kilowatt is equal to which of the following?

   a. 1 W

   b. 10 W

   c. 100 W

   d. 1,000 W

Power Ratings

8. What is the rate at which an electrical device uses energy called?

   a. energy rating

   b. power rating

   c. star rating

   d. electrical rating

| Directed Reading B *continued*

Circle the letter of the best answer for each question.

9. What will tell you the power rating of an electrical device?

a. the electric company

b. the label

c. the EPA

d. the voltage

MEASURING ELECTRICAL ENERGY

10. What is used to find how much a home has to pay for electric power?

a. power and temperature

b. time and temperature

c. power and time

d. current and power

11. Which of the following is the equation for electrical energy?

a. $E = P \times t$

b. $V = C + P$

c. $P = E \times I$

d. $W = P \times V$

Measuring Household Energy Use

12. What is a unit used to label electrical energy?

a. amps

b. kilowatt-hours

c. volts

d. ohms

| Directed Reading B *continued*

Circle the letter of the best answer for each question.

13. What do electric companies use to find out how many hours of energy are used?

   a. phone lines

   b. volts

   c. photocells

   d. meters

How to Save Energy

14. Which of the following is NOT a way to save energy?

   a. turning off the TV

   b. leaving the lights on

   c. running a fan to stay cool

   d. replacing high power items

It's All About the Bulb

15. How do incandescent light bulbs compare to fluorescent light bulbs?

   a. more light/more energy

   b. more light/less energy

   c. same light/more energy

   d. same light/less energy

Energy-Saving Programs

16. What program promotes the efficient use of energy?

   a. the ENERGY STAR® program

   b. the ENERGY LABEL® program

   c. the ENERGY EFFICIENCY® program

   d. the ENERGY AWARENESS® program

Directed Reading B

Section: Electric Circuits

Circle the letter of the best answer for each question.

1. What is the closed path called where the start and end point are the same?

   a. circuit

   b. load

   c. fuse

   d. battery

PARTS OF AN ELECTRIC CIRCUIT

2. What is a complete, closed path where electric charges flow?

   a. electric charge

   b. electric circuit

   c. electric current

   d. electric meter

3. What are the three basic parts of a circuit?

   a. energy source, wires, battery

   b. energy source, wires, load

   c. energy source, load, battery

   d. load, switch, wires

4. Which of the following are examples of an energy source?

   a. wires and currents

   b. volts and amps

   c. fuses and switches

   d. batteries and photocells

5. Which of the following are examples of loads?

   a. appliances and motors

   b. electrolytes and electrodes

   c. photocells and thermocouples

   d. resistance and power

A Switch To Control a Circuit
Circle the letter of the best answer for each question.

6. What is used to open and close a circuit?

  a. button

  b. string

  c. switch

  d. loop

TYPES OF CIRCUITS

7. Which of the following are the two types of circuits?

  a. opened and closed

  b. on and off

  c. charged and uncharged

  d. series and parallel

Series Circuits

8. How are all parts of a series circuit connected?

  a. in a single loop

  b. in a triple loop

  c. in a double loop

  d. in an open loop

9. What do all loads in a series circuit share?

  a. the same voltage

  b. the same current

  c. the same wires

  d. the same resistance

10. What happens in a series circuit if more bulbs are added?

  a. The current goes up.

  b. The current drops.

  c. The current and resistance stay the same.

  d. The resistance drops.

Uses for Series Circuits
Circle the letter of the best answer for each question.

11. How many pathways for moving charges are there in a series circuit?

   a. 1   **c.** 3

   b. 2   **d.** 4

12. What happens if there is a break in a series circuit?

   a. The charges slow down.

   b. The charges stop flowing.

   c. The charges speed up.

   d. The charges stay the same.

13. Why would you NOT want to wire your house using only series circuits?

   a. nothing would work

   b. nothing could be turned on

   c. if one thing won't work, other things won't work

   d. nothing could be turned off

Parallel Circuits

14. How are loads connected in a parallel circuit?

   a. side by side

   b. on top of each other

   c. under each other

   d. crossing each other

Uses for Parallel Circuits

15. What happens in a parallel circuit if one load is broken?

   a. The charges stop flowing.

   b. The charges still flow.

   c. The temperature rises.

   d. The circuit breaker turns off.

| Directed Reading B *continued*

Read the words in the box. Read the sentences. <u>Fill in each blank</u> with the word that best completes the sentence.

| series | parallel | loads | voltage | switch |

16. A circuit where loads are connected side by side is a

_________________________________ circuit.

17. Objects that use electrical energy are all

_________________________________ in a large circuit.

18. A circuit where all parts are connected in a single loop is

a _________________________ circuit.

19. Each bulb in a parallel circuit uses the full

_________________________________ of a battery.

20. Each outlet in a home has its own _____________________________.

HOUSEHOLD CIRCUIT SAFETY

<u>Circle the letter</u> of the best answer for each question.

21. In a home, where do circuits branch out from?

a. the wall

b. the floor

c. a fuse box

d. an outlet

Circuit Failure

22. What could cause circuits to fail?

a. if they are underloaded

b. if they are overloaded

c. if they are left open

d. if they are turned up

| Directed Reading B *continued*

Fuses

Circle the letter of the best answer for each question.

23. What happens to the metal strip in a fuse if the current is too high?

    a. It gets bigger.

    b. It melts.

    c. It gets smaller.

    d. It explodes.

Circuit Breakers

24. What do you call a switch that automatically opens if the current is too high?

    a. fuse

    b. circuit breaker

    c. switch

    d. series circuit

25. Which of the following happens when a circuit breaker opens?

    a. charges stop flowing

    b. charges start flowing

    c. current increases

    d. switch closes

Electrical Safety Tips

26. Which of the following is NOT a way to stay safe while using electrical energy?

    a. By making sure insulation on cords is not worn.

    b. By not overloading circuits.

    c. By using electrical devices with wet hands.

    d. By never putting objects other than a plug in an outlet.

Vocabulary and Section Summary

Electric Charge and Static Electricity

VOCABULARY

In your own words, write a definition of the following terms in the space provided.

1. law of electric charges

2. electric force

3. electric field

4. electrical conductor

5. electrical insulator

6. static electricity

7. electric discharge

▎Vocabulary and Section Summary *continued*

SECTION SUMMARY

Read the following section summary.

- The law of electric charges states that like charges repel and opposite charges attract.
- The size of the electric force between two objects depends on the size of the charges exerting the force and the distance between the objects.
- Charged objects exert a force on each other and can cause each other to move.
- Objects become charged when they gain or lose electrons.
- Objects may become charged by friction, conduction, or induction.
- Charges are not created or destroyed and are said to be conserved.
- Charges move easily in conductors but do not move easily in insulators.
- Static electricity is the buildup of electric charges on an object. It is lost through electric discharge.

Vocabulary and Section Summary

Electric Current and Electrical Energy
VOCABULARY

In your own words, write a definition of the following terms in the space provided.

1. electric current

____

____

2. voltage

____

____

3. resistance

____

____

4. cell

____

____

5. thermocouple

____

____

6. photocell

____

____

▌Vocabulary and Section Summary *continued*

SECTION SUMMARY

Read the following section summary.

- Electric current is the rate at which charges pass a given point.

- An electric current can be made when there is a potential difference between two points.

- As voltage, or potential difference increases, current increases.

- An object's resistance varies depending on the object's material, thickness, length, and temperature. As resistance increases, current decreases.

- Cells and batteries convert chemical energy or radiant energy into electrical energy.

- Thermocouples and photocells are devices used to generate electrical energy.

Vocabulary and Section Summary

Electrical Calculations

VOCABULARY

In your own words, write a definition of the following term in the space provided.

1. electric power

____

____

SECTION SUMMARY

Read the following section summary.

- Ohm's law describes the relationship between current, resistance, and voltage.

- Electric power is the rate at which electrical energy is changed into other forms of energy.

- The power rating of an electrical device describes the amount of energy the device uses.

- Electrical energy is electric power multiplied by time. It is usually expressed in kilowatt-hours.

- Using energy-efficient appliances and changing the ways we use certain electrical appliances can save energy.

Vocabulary and Section Summary

Electric Circuits

VOCABULARY

In your own words, write a definition of the following terms in the space provided.

1. series circuit

____

____

2. parallel circuit

____

____

SECTION SUMMARY

Read the following section summary.

- Circuits consist of an energy source, a load, wires, and, in some cases, a switch.

- All parts of a series circuit are connected in a single loop. The loads in a parallel circuit are on separate branches.

- Circuits fail through a short circuit or an overload. Fuses or circuit breakers protect against circuit failure.

- It is important to follow safety tips when using electrical energy.

Section Review

Electric Charge and Static Electricity

USING KEY TERMS

For each pair of terms, explain how the meanings of the terms differ.

1. *static electricity* and *electric discharge*

2. *electric force* and *electric field*

3. *electrical conductor* and *electrical insulator*

UNDERSTANDING KEY IDEAS

______ **4.** Which of the following is an insulator?
- **a.** copper
- **b.** rubber
- **c.** aluminum
- **d.** iron

5. Compare the three methods of charging.

6. What does the law of electric charges say about two objects that are positively charged?

7. Give two examples of static electricity.

8. List two examples of electric discharge.

____

____

CRITICAL THINKING

9. Analyzing Processes Imagine that you touch the top of an electroscope with an object. The metal leaves spread apart. Can you determine whether the charge is positive or negative? Explain your answer.

____

____

____

____

10. Applying Concepts Why is it important to touch a charged object to the metal rod of an electroscope and not to the rubber stopper?

____

____

____

____

| Section Review *continued*

INTERPRETING GRAPHICS

The image below shows two charged balloons. Use the image below to answer the questions that follow.

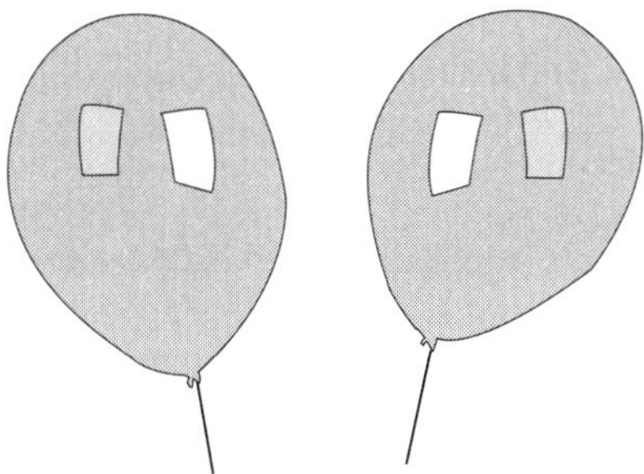

11. Do the balloons have the same charge or opposite charges? Explain your answer.

____

____

____

____

12. How would the image look if each balloon were given the charge opposite to the charge it has now? Explain your answer.

____

____

____

____

____

Section Review

Electric Current and Electrical Energy
USING KEY TERMS

Complete each of the following sentences by choosing the correct term from the word bank.

voltage   electric current   resistance   cell

1. The rate at which charges pass a point is a(n) _______________________.

2. The opposition to the flow of charge is _______________________.

3. Another term for *potential difference* is _______________________.

4. A device that changes chemical energy into electrical energy is a(n)

_______________________.

UNDERSTANDING KEY IDEAS

_______ **5.** Which of the following factors affects the resistance of an object?
   a. thickness of the object
   b. length of the object
   c. temperature of the object
   d. All of the above

6. Name the parts of a cell, and explain how they work together to produce an electric current.

____

____

7. Compare alternating current with direct current.

____

____

8. How do the currents produced by a 1.5 V flashlight cell and a 12 V car battery compare if the resistance is the same?

____

____

9. How does increasing the resistance affect the current?

____

____

CRITICAL THINKING

10. Making Comparisons A friend is having trouble studying the types of cells in this section. Explain to your friend how the terms *photocell* and *thermocouple* hold clues that can help him or her remember the type of energy taken in by each device.

____

____

____

11. Making Inferences Why do you think some calculators that contain photocells also contain batteries?

____

____

____

12. Applying Concepts Which wire would have the lowest resistance: a long, thin iron wire at a high temperature or a short, thick copper wire at a low temperature?

____

____

____

INTERPRETING GRAPHICS

13. The wires shown below are made of copper and have the same temperature. Which wire should have the lower resistance? Explain your answer.

____

____

____

Section Review

Electrical Calculations

USING KEY TERMS

1. In your own words, write a definition for the term *electric power.*

____

____

UNDERSTANDING KEY IDEAS

______ **2.** Which of the following is Ohm's law?
- **a.** $E = P \times t$
- **b.** $I = V \times R$
- **c.** $P = V \times I$
- **d.** $V = I \times R$

3. Circuit A has twice the resistance of circuit B. The voltage is the same in each circuit. Which circuit has the higher current?

____

____

______ **4.** The power rating of an electrical device describes
- **a.** the rate at which it gives off thermal energy.
- **b.** the rate at which it uses solar energy.
- **c.** the rate at which it uses electrical energy.
- **d.** the rate at which it saves energy.

5. Describe two ways that you can save energy at home.

____

____

6. Name one energy-saving electrical device that is available today.

____

____

Section Review *continued*

MATH SKILLS

7. Use Ohm's law to find the voltage needed to make a current of 3 A in a resistance of 9 Ω. Show your work below.

8. How much electrical energy does a 40 W light bulb use if it is left on for 12 h? Show your work below.

CRITICAL THINKING

9. Applying Concepts Explain why increasing the voltage applied to a wire can have the same effect on the current in the wire that decreasing the resistance of the wire does.

____

____

____

____

10. Identifying Relationships Using the equations in this section, develop an equation to find electrical energy from time, current, and resistance.

____

____

____

____

Section Review

Electric Circuits

USING KEY TERMS

1. In your own words, write a definition for each of the following terms: *series circuit* and *parallel circuit*.

____

____

____

UNDERSTANDING KEY IDEAS

_________ **2.** Which part of a circuit changes electrical energy into another form of energy?

    a. energy source

    b. wire

    c. switch

    d. load

3. Name and describe the three essential parts of a circuit.

____

____

____

4. How do fuses and circuit breakers protect your home against electrical fires?

____

____

CRITICAL THINKING

5. Forming Hypotheses Suppose that you turn on the heater in your room and all of the lights in your room go out. Propose a reason why the lights went out.

____

____

____

6. Applying Concepts Will a fuse work successfully if it is connected in parallel with the device it is supposed to protect? Explain your answer.

____

____

____

____

INTERPRETING GRAPHICS

7. Look at the circuits below. Identify each circuit as a parallel circuit or a series circuit.

a.    **b.** 

a. ____

b. ____

Chapter Review

USING KEY TERMS

The statements below are false. For each statement, replace the underlined term to make a true statement.

1. Charges flow easily in an <u>electrical insulator</u>.

____

2. Lightning is a form of <u>static electricity</u>.

____

3. A <u>thermocouple</u> converts chemical energy into electrical energy.

____

4. <u>Voltage</u> is the opposition to the current by a material.

____

5. <u>Electric force</u> is the rate at which electrical energy is converted into other forms of energy.

____

6. Each load in a <u>parallel circuit</u> has the same current.

____

UNDERSTANDING KEY IDEAS

Multiple Choice

_______ **7.** Two objects repel each other. What charges might the objects have?
    a. positive and positive   **c.** negative and negative
    b. positive and negative   **d.** Both (a) and (c)

_______ **8.** Which device converts chemical energy into electrical energy?
    a. lightning rod   **c.** light bulb
    b. cell   **d.** switch

_______ **9.** Which of the following wires has the lowest resistance?
    a. a short, thick copper wire at 25°C
    b. a long, thick copper wire at 35°C
    c. a long, thin copper wire at 35°C
    d. a short, thick iron wire at 25°C

_______ **10.** An object becomes charged when the atoms in the object gain or lose
    a. protons.   **c.** electrons.
    b. neutrons.   **d.** All of the above

| Chapter Review *continued*

_______**11.** Which of the following devices does NOT protect you from electrical fires?
  a. electric meter
  b. circuit breaker
  c. fuse
  d. ground fault circuit interrupter

_______**12.** For a cell to produce a current, the electrodes of the cell must
  a. have a potential difference.
  b. be in a liquid.
  c. be exposed to light.
  d. be at two different temperatures.

_______**13.** The outlets in your home provide
  a. direct current.
  b. alternating current.
  c. electric discharge.
  d. static electricity.

Short Answer

14. Describe how a switch controls a circuit.

____

____

15. Name the two factors that affect the strength of electric force, and explain how they affect electric force.

____

____

16. Describe how direct current differs from alternating current.

____

____

17. What is a power rating? Name one electrical device that has a high power rating.

____

Math Skills

18. What voltage is needed to produce a 6 A current in an object that has a resistance of 3 Ω? Show your work below.

19. Find the current produced when a voltage of 60 V is applied to a resistance of 15 Ω. Show your work below.

20. What is the resistance of an object if a voltage of 40 V produces a current of 5 A? Show your work below.

21. A light bulb is rated at 150 W. How much current is in the bulb if 120 V is applied to the bulb? Show your work below.

| Chapter Review *continued*

CRITICAL THINKING

22. Concept Mapping Use the following terms to create a concept map: *electric current, battery, charges, photocell, thermocouple, circuit, parallel circuit,* and *series circuit.*

| Chapter Review *continued*

23. Making Inferences Suppose your science classroom was rewired over the weekend. On Monday, you notice that the lights in the room must be on for the fish-tank bubbler to work. And if you want to use the computer, you must turn on the overhead projector. Describe what mistake the electrician made when working on the circuits in your classroom.

____

____

____

24. Applying Concepts You can make a cell by using an apple, a strip of copper, and a strip of silver. Explain how you would construct the cell, and identify the parts of the cell. What type of cell did you make? Explain your answer.

____

____

____

25. Applying Concepts Your friend shows you a magic trick. First, she rubs a plastic pipe on a piece of wool. Then, she holds the pipe close to an empty soda can that is lying on its side. When the pipe is close to the can, the can rolls toward the pipe. Explain how this trick works.

____

____

____

INTERPRETING GRAPHICS

26. Classify the objects in the image below as electrical conductors or electrical insulators.

____

____

____

Reinforcement

Charge!

Complete this worksheet after you have finished reading the section "Electric Charge and Static Electricity."

There are three ways for an object to gain a charge: friction, conduction, and induction. When it loses its charge, it experiences electric discharge. Label the following pictures as examples of *conduction, induction, friction,* or *electric discharge*.

1.

____

4.

____

2.

____

5.

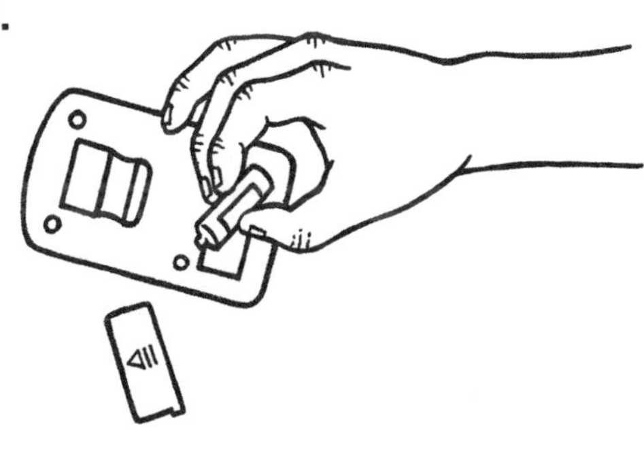

____

3.

____

6.

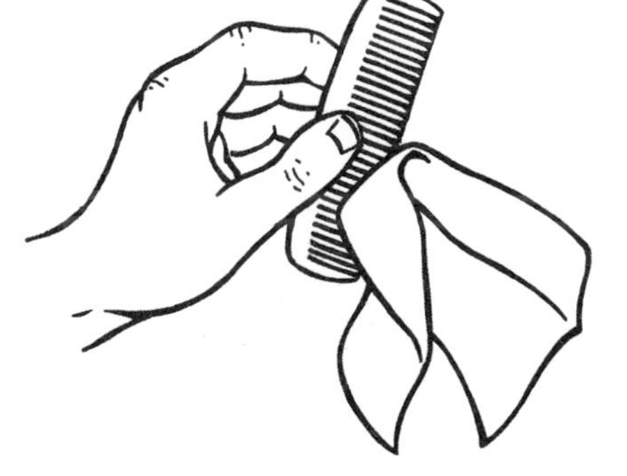

____

Name ______________________ Class ______________ Date ______________

Skills Worksheet

Reinforcement

Electric Circuits

Complete this worksheet after you have finished reading the section "Electric Circuits."

Two electric circuits powered by cells are shown below. Answer the following questions based on the information given in the diagrams. Questions 1–6 refer to Figure 1, and Questions 8–12 refer to Figure 2.

Label the parts of the circuit and the cell by writing the letter that corresponds to the appropriate part in the space provided.

Figure 1

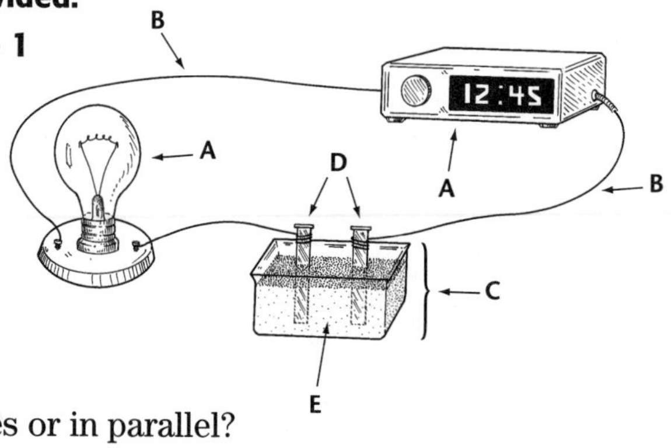

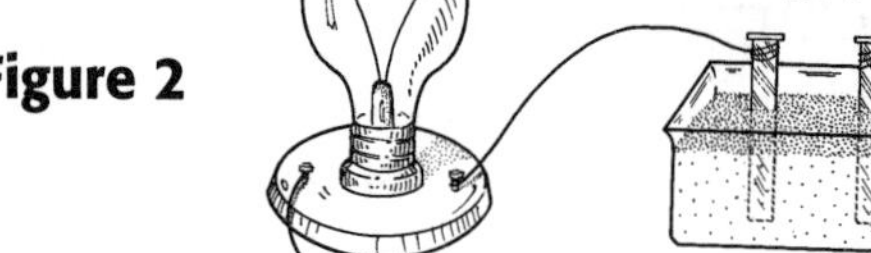

______ **1.** load

______ **2.** electrode

______ **3.** wire

______ **4.** electrolyte

______ **5.** energy source

6. Is this circuit connected in a series or in parallel?

____

7. A cell that contains liquid electrolytes is called a(n) ______________ cell.

Figure 2

8. What is the power in this circuit? ______________

9. What is the voltage in this circuit? ______________

10. Recall that $I = P/V$. If you divide the power of the circuit by its voltage, you'll find the circuit's current. What is the current of this circuit?

____

11. Remember that Ohm's law can be rearranged to say: $R = V/I$. If you divide the circuit's voltage by its current, you'll get the resistance of the circuit. What is the resistance caused by the light bulb?

____

12. This cell contains a solid electrolyte, so it is a(n) ______________ cell.

Critical Thinking

Potentially Shocking

Stuart, FL – Four bald eagles and a pelican in search of a perch were electrocuted when they landed atop two power poles and were jolted with 13,000 volts of electricity. Now utility and game officials are taking steps to ensure no more of the regal birds are injured. The poles, located beside the 21,000 acre DuPuis Wildlife and Recreation Area southwest of Stuart, now have a safety platform, and the wires were moved below the crossbar. It's not known exactly how the eagles came in contact with the electrical wires. It could have happened when an eagle's 6- to 8-foot wing touched a wire carrying a current while the other wing touched the pole's (grounded) metal crossbar, thus becoming a conduit for electricity. Or the eagle could have stepped on a hot wire with one foot and the metal crossbar with the other. Electrocution is "one of the major causes of death" for ospreys and bald eagles.

From "Four bald eagles, pelican electrocuted on poles" from *The Palm Beach Post*. Copyright © 1998 by **The Palm Beach Post**. Reprinted with permission of the copyright holder.

ASSESSING INFORMATION

1. Why is it dangerous to have power poles located next to a wildlife area?

____

____

____

DEMONSTRATING REASONED JUDGMENT

2. Salt water is a good conductor of electric current. Like our blood, birds' blood is made mostly of water and salt. How does the composition of the birds' blood affect their resistance?

HELPFUL HINT
Remember Ohm's Law

____

____

Critical Thinking *continued*

ESTABLISHING CONNECTIONS

3. What is the voltage of the metal crossbar?

____

____

USEFUL TERM
"hot" electrically charged with a high-voltage current

DEMONSTRATING REASONED JUDGMENT

4. Why would an eagle be electrocuted when touching a "hot" wire and the metal crossbar at the same time?

____

____

____

5. The power line carries 13,000 V along all points of the wire. Calculate the potential difference between the feet of a small bird perched on this power line.

____

____

____

DRAWING CONCLUSIONS

6. We frequently see small birds perched on a power line. Why aren't they electrocuted?

____

____

____

Section Quiz

Section: Electric Charge and Static Electricity

Write the letter of the correct answer in the space provided.

______ **1.** What are the tiny particles that make up matter?
    a. charges   **c.** electricity
    b. atoms   **d.** forces

______ **2.** What is the region around a charged object where an electric force is present?
    a. electric force   **c.** electric field
    b. proton   **d.** electron

______ **3.** Which part of the atom has a positive charge?
    a. proton   **c.** neutron
    b. electron   **d.** nucleus

______ **4.** What happens when electrons move from one object to another by direct contact?
    a. friction   **c.** induction
    b. detection   **d.** conduction

______ **5.** What would you use to see if something is charged?
    a. electric reader   **c.** electroscope
    b. thermometer   **d.** electric force

______ **6.** If a material does not allow charges to move through it easily, what is it called?
    a. electrical insulator   **c.** static electricity
    b. electrical conductor   **d.** electric discharge

______ **7.** What is the loss of static electricity as charges move off an object?
    a. static   **c.** friction
    b. electric discharge   **d.** induction

Section Quiz

Section: Electric Current and Electrical Energy

Write the letter of the correct answer in the space provided.

_______ **1.** What does the size of a current depend upon?
    a. protons   **c.** voltage
    b. electrons   **d.** charge

_______ **2.** What do you call materials with a resistance of 0 Ω?
    a. conductors   **c.** resistors
    b. superconductors   **d.** photocells

_______ **3.** What is the part of the cell where charges enter or exit?
    a. nucleus   **c.** electron
    b. electrolyte   **d.** electrode

Match the correct definition with the correct term. Write the letter in the space provided.

_______ **4.** when charges shift from flowing from one direction to another

_______ **5.** the potential difference between two points in a circuit

_______ **6.** the rate at which charges pass through a given point

_______ **7.** converts light energy into electrical energy

_______ **8.** converts thermal energy into electrical energy

_______ **9.** the opposition to the flow of electric charge

_______ **10.** changes chemical or radiant energy into electrical energy

a. cell

b. electric current

c. photocell

d. voltage

e. alternating current

f. thermocouple

g. resistance

Section Quiz

Section: Electrical Calculations

Write the letter of the correct answer in the space provided.

_______ **1.** As resistance goes up, what happens to the current?
   a. Current goes up.
   b. Current goes down.
   c. Current stays the same.
   d. Current disappears.

_______ **2.** What is the rate at which electrical energy is changed into other forms
   of energy?
   a. electric power
   b. voltage
   c. electrical energy
   d. amps

_______ **3.** What is the voltage if the current is 4 A and the resistance is 10 Ω?
   a. 4 V
   b. 40 R
   c. 10 V
   d. 40 V

_______ **4.** How much electrical energy does a 75 W light bulb use if it is on
   for 4 hours?
   a. 150 kWh
   b. 150 W
   c. 300 kWh
   d. 300 W

_______ **5.** In the formula $P = V \times I$, what does the I stand for?
   a. current
   b. intensity
   c. voltage
   d. resistance

_______ **6.** The power rating of an electrical device is the
   a. amount of energy the device uses.
   b. amount of time a device can be used.
   c. number of watts a device uses.
   d. rate at which the device uses energy.

Section Quiz

Section: Electric Circuits

Write the letter of the correct answer in the space provided.

_______ **1.** Circuits need three basic parts, an energy source, wires, and
what else?
    a. charge   **c.** load
    b. force   **d.** energy

_______ **2.** How many pathways are there for moving charges in a series circuit?
    a. one   **c.** three
    b. two   **d.** four

_______ **3.** In a short circuit, as the resistance decreases, what happens to
the current?
    a. increases   **c.** does not change
    b. decreases   **d.** drops to 0 A

Match the correct definition with the correct term. Write the letter in the space provided.

_______ **4.** a circuit in which loads are connected side   **a.** switch
by side
    b. parallel circuit

_______ **5.** a complete, closed path through which   **c.** series circuit
electric charges flow
    d. electric circuit

_______ **6.** a device used to open and close a circuit

_______ **7.** a circuit in which all parts are connected in
a single loop

Introduction to Electricity

MULTIPLE CHOICE

Write the letter of the correct answer in the space provided.

_______ **1.** What are the small particles that make up matter?
- **a.** charges
- **b.** electrons
- **c.** atoms
- **d.** protons

_______ **2.** Why are protons and electrons attracted to each other?
- **a.** because protons have a positive charge and electrons have a negative charge
- **b.** because protons have a negative charge and electrons have a positive charge
- **c.** because they both are positively charged
- **d.** because they both are negatively charged

_______ **3.** Why is it important that the electrons and protons are attracted to each other?
- **a.** The attraction determines the size of the atom.
- **b.** The attraction keeps the electrons from flying away from the nucleus.
- **c.** The attraction determines the size of the nucleus.
- **d.** The attraction determines the charge of the atom.

_______ **4.** The size of an electric force depends upon which two things?
- **a.** the amount of each charge and the size of the electric field
- **b.** the distance between the charges and the size of the electric field
- **c.** the number of protons and the distance between the charges
- **d.** the amount of each charge and the distance between the charges

_______ **5.** What method is involved when charges in an uncharged metal object are rearranged without direct contact with a charged object?
- **a.** friction
- **b.** induction
- **c.** convection
- **d.** conduction

_______ **6.** Which of the following is positively charged?
- **a.** protons
- **b.** neutrons
- **c.** electrons
- **d.** atoms

_______ **7.** Which of the following is NOT an insulator?
- **a.** air
- **b.** wood
- **c.** glass
- **d.** copper

MATCHING

Match the correct description with the correct term. Write the letter in the space provided.

_______ **8.** The higher this is, the lower the current.

_______ **9.** This device uses the temperature difference in wires to convert thermal energy into electrical energy.

_______ **10.** As this increases, so does the current.

_______ **11.** This method of charging happens when you rub a balloon on your hair.

_______ **12.** This is a material, like metal, that allows charges to move easily.

_______ **13.** This is the region around a charged object where a force is exerted on other objects.

_______ **14.** These are more efficient than incandescent light bulbs.

_______ **15.** This is expressed in amps.

_______ **16.** This is the force between two charged objects.

_______ **17.** Lightning is an example of this.

a. resistance

b. electrical conductor

c. friction

d. voltage

e. electric discharge

f. electric current

g. electric force

h. electric field

i. fluorescent bulbs

j. thermocouple

| Chapter Test A *continued*

MULTIPLE CHOICE

Write the letter of the correct answer in the space provided.

_______**18.** What generates electrical energy from chemical energy?
    a. cell   **c.** circuit
    b. switch   **d.** current

_______**19.** What is the voltage if the current is 0.4 A and the resistance is 3 Ω?
    a. 12 V   **c.** 1.2 V
    b. 12 Ω **d.** 1.2 Ω

_______**20.** A video monitor draws 1.5 A at a voltage of 150 V. What is the power
    rating of the monitor?
    a. 220 W   **c.** 150 W
    b. 225 W   **d.** 2,250 W

_______**21.** How much electrical energy does a 75 W light bulb use if it is on
    for 5 hours?
    a. 375 W   **c.** 300 W
    b. 375 kWh   **d.** 300 kWh

_______**22.** What is the third part of an electric circuit besides the wires and
    the load?
    a. force   **c.** current
    b. voltage   **d.** energy source

_______**23.** What is a switch that automatically opens if the current is too high?
    a. fuse   **c.** circuit breaker
    b. conductor   **d.** insulator

MATCHING

Match each drawing to the correct term. Write the letter of the correct answer in the space provided.

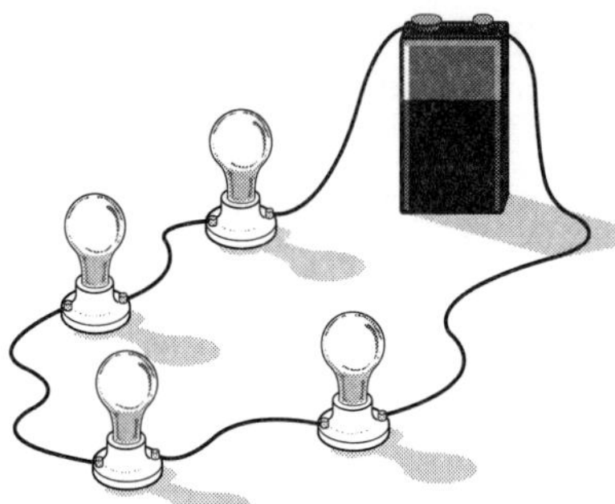

a.

b.

_______**24.** parallel circuit

_______**25.** series circuit

Chapter Test B

Introduction to Electricity

USING KEY TERMS

Use the terms from the following list to complete the sentences below. Each term may be used only once. Some terms may not be used.

electric current	series circuit	electrical insulators
electric power	electrical	voltage
static electricity		

1. Plastic, glass, wood, and air are examples of good ________________________.

2. Electrons moving in a wire make up ____________________ and provide

energy to the things that you use each day.

3. Burglar alarms are best wired using a ____________________.

4. When the voltage is in volts and the current is in amperes,

____________________ is expressed in watts.

5. When your clothes come out of the dryer stuck together, they are full of

____________________.

UNDERSTANDING KEY IDEAS

Write the letter of the correct answer in the space provided.

______ **6.** Which of these would lower the electrical resistance of a wire?
    a. making the wire thinner
    b. increasing the wire's length
    c. lowering the temperature of the wire
    d. using denser material for the wire

______ **7.** What happens if you rub a glass rod with a piece of silk and the rod
becomes positively charged?
    a. Electrons on the rod are destroyed.
    b. The silk becomes negatively charged.
    c. Protons move to the rod.
    d. The glass attracts more protons.

______ **8.** When you flip the switch on a flashlight, what is immediately set up?
    a. electric current   **c.** electric field
    b. voltage   **d.** resistance

______ **9.** What does the amount of energy released per charge depend upon?
    a. voltage   **c.** current
    b. resistance   **d.** friction

______ **10.** In a photocell, what gains energy to move between atoms?
    a. neutrons   **c.** protons
    b. electrons   **d.** atoms

11. What does Ohm's law tell us?

____

____

____

12. Why is a fluorescent light bulb more efficient than an incandescent light bulb?

____

____

____

13. How does a circuit breaker help to protect against short circuits and circuit overloads?

____

____

____

14. A string of lights wired together in a series has a burned out bulb. Why do all of the lights go out?

____

____

____

15. Find the resistance of a circuit that draws a 1.5 A when 3.0 V are applied. Show your work.

CRITICAL THINKING

16. Why would it be dangerous to replace a blown fuse with a copper penny?

____

____

____

____

17. As an added safety precaution, would it be useful to insert two fuses in a circuit instead of one? Explain your answer.

____

____

____

____

18. Why could standing on a beach or in the open on a golf course make your body act like a lightning rod?

____

____

____

| Chapter Test B *continued*

INTERPRETING GRAPHICS

Examine the diagram below and answer the question that follows.

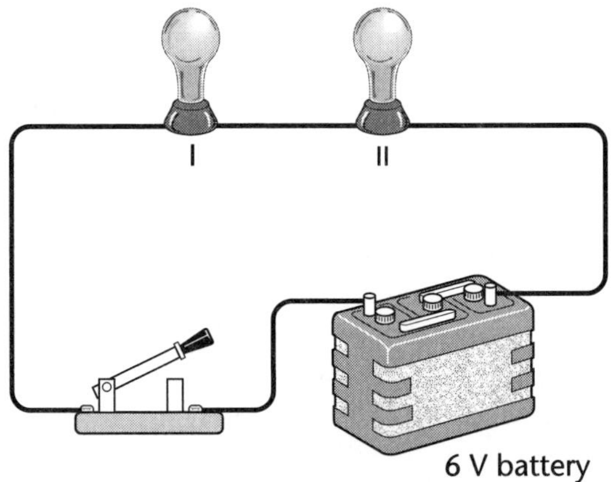

19. If one bulb in this circuit burns out, how will the other bulb be affected? Explain your answer.

____

____

____

Chapter Test C

Introduction to Electricity
MULTIPLE CHOICE
Circle the letter of the best answer for each question.

1. What is the rate at which charges pass a given point?

   a. electrical energy

   b. ohms

   c. electric current

   d. voltage

2. What goes down as resistance goes up?

   a. Power goes down.

   b. Charge goes down.

   c. Current goes down.

   d. Temperature goes down.

3. How much energy do fluorescent light bulbs use compared to incandescent light bulbs?

   a. Fluorescent bulbs use more energy.

   b. Fluorescent bulbs use less energy.

   c. Fluorescent bulbs use the same energy.

   d. They can't be compared.

4. What is the switch called that opens if the current is too high?

   a. circuit breaker

   b. conductor

   c. fuse

   d. insulator

5. What is the equation for electrical energy?

   a. $E = P \times t$ **c.** $E = P + t$

   b. $E = P \div t$ **d.** $E = P - t$

MATCHING

Read the description. Then, <u>draw a line</u> from the dot next to each description to the matching word.

6. charging by wiping electrons from one object to another object ●

7. rearranging charges in an uncharged metal object without direct contact with a charged object ●

8. charging when electrons move by direct contact ●

9. losing static electricity as charges move off an object ●

a. conduction

b. induction

c. friction

d. electric discharge

---

10. a material in which charges can move freely ●

11. a material in which charges cannot move freely ●

12. the rate at which electrical energy is converted into forms of other energy ●

a. electrical insulator

b. electric power

c. electrical conductor

| Chapter Test C *continued*

FILL-IN-THE-BLANK

Read the words in the box. Read the sentences. <u>Fill in each blank</u> with the word or phrase that best completes the sentence.

photocell	electrolyte	electrode	resistance

13. A mixture of chemicals in a cell is a(n)

_______________________.

14. The opposition to the flow of electric charge is

_______________________.

15. A device that converts light energy into electrical energy is called

a(n) _______________________.

16. The part of the cell through which charges enter and exit is the

_______________________.

law of electric charges	thermocouple
energy source	

17. The law that states that like charges repel and opposite charges

attract is called the _______________________.

18. A device that converts thermal energy into electrical energy is

called a(n) _______________________.

19. The three basic parts of a circuit are wires, load, and a(n)

_______________________.

MATCHING

Read the description. Then, <u>draw a line</u> from the dot next to each description to the matching picture.

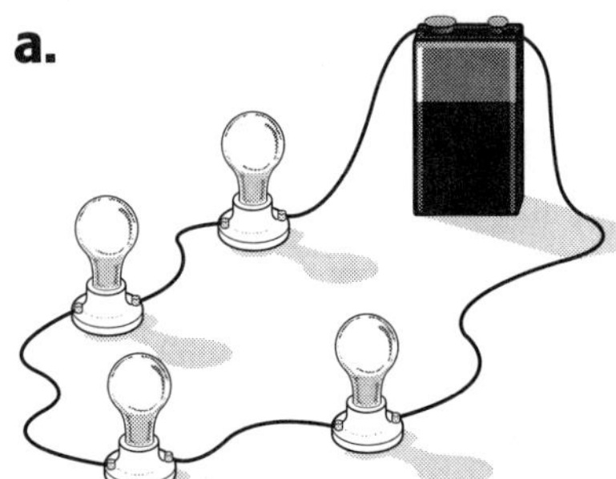

a.

20. parallel circuit ●

21. series circuit ●

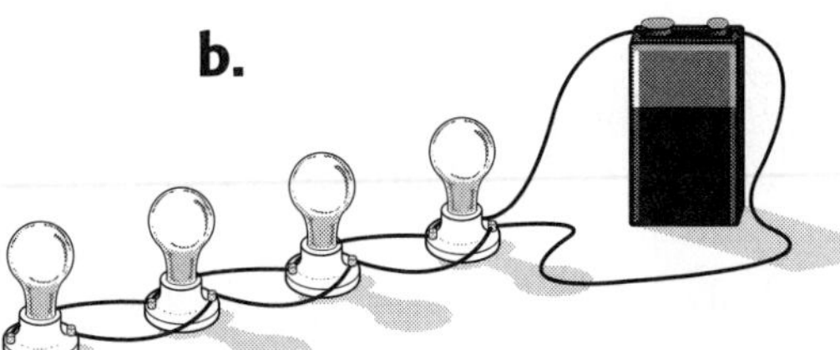

b.

  SKILLS PRACTICE

Performance-Based Assessment

OBJECTIVE

In this activity, you will connect resistors in series and in parallel and note how each arrangement affects the voltage readings in simple electric circuits.

KNOW THE SCORE!

As you work through the activity, keep in mind that you will be earning a grade for the following:

- how well you work with the materials and choose the correct scenario (20%)
- how well you explain your observations (40%).
- whether you construct an operational model (40%)

Using Scientific Methods

ASK A QUESTION

How does a parallel circuit differ from a series circuit?

MATERIALS AND EQUIPMENT

- Broomstick or wooden dowel
- 6 V battery
- 2 wires with alligator clips
- 1,000 ohm resistors (3)
- voltmeter

SAFETY INFORMATION

Keeping your work area clean helps maintain a safe work environment.

FORM A HYPOTHESIS

You will be attaching resistors to terminals of the battery. How do you think this will affect the voltage?

TEST THE HYPOTHESIS

1. Attach three resistors between the terminals of the battery as shown in diagram A at left. The resistors are now *in series*: all of the current in the circuit flows through each resistor.

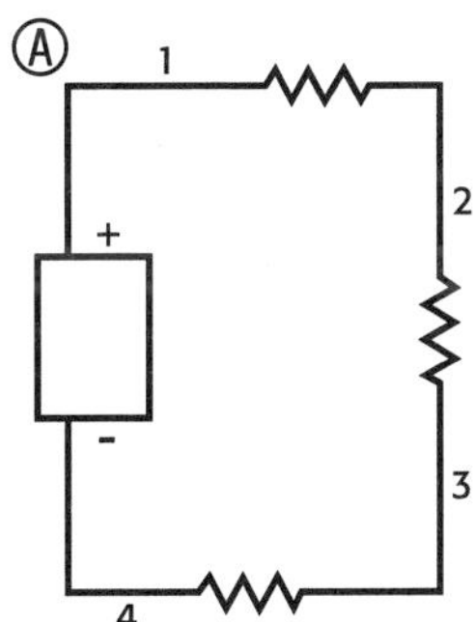

| Performance-Based Assessment *continued*

---

2. Attach the black probe of the voltmeter to the negative terminal of the battery. With the red probe, measure the voltage in the circuit at each of the four points shown in the diagram. Record your data below.

____

____

3. Disassemble the circuit. Now, attach three resistors between the terminals of the battery as shown in diagram B. The two resistors between points 2 and 3 are now *in parallel*: half the circuit current passes through each of them. The entire current still passes through the resistor between points 1 and 2.

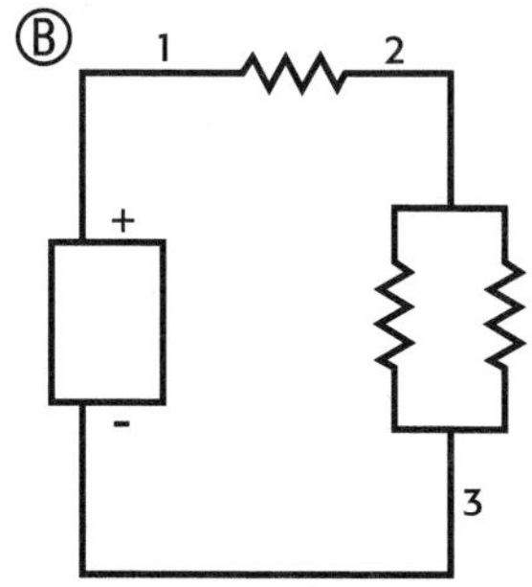

4. Attach the black probe of the voltmeter to the negative terminal of the battery. With the red probe, measure the voltage in the circuit at each of the three points shown in the diagram. Record your data below.

____

____

ANALYZE THE RESULTS

5. Compare the voltage drop between points 1 and 2 with the voltage drop between points 2 and 4 when the resistors are in series. Explain what is happening.

____

____

____

____

6. Compare the drop in voltage between points 1 and 2 with the drop between points 2 and 3 when the resistors are in parallel.

____

____

____

Performance-Based Assessment *continued*

7. What is the overall resistance between points 2 and 3 when the resistors are in parallel?

____

____

____

DRAW CONCLUSIONS

8. What differences do you see in your results? Why do you think these differences occurred?

____

____

____

Standardized Test Preparation

READING

Read each of the passages below. Then, answer the questions that follow each passage.

Passage 1 In 1888, Frank J. Sprague developed a way to operate trolleys by using electrical energy. These electric trolleys ran on a metal track and were connected by a pole to an overhead power line. Electric charges flowed down the pole to motors in the trolley. A wheel at the top of the pole, called a <u>shoe</u>, rolled along the power line and allowed the trolley to move along its track without losing contact with its source of electrical energy. The charges passed through the motor and then returned to a generator by way of the metal track.

_______ **1.** In this passage, what does the word *shoe* mean?
- **A** a type of covering that you wear on your foot
- **B** a device that allowed a trolley to get electrical energy
- **C** a flat, U-shaped metal plate nailed to a horse's hoof
- **D** the metal track on which trolleys ran

_______ **2.** What is the main purpose of this passage?
- **F** to inform the reader
- **G** to influence the reader's opinion
- **H** to express the author's opinion
- **I** to make the reader laugh

_______ **3.** Which of the following statements describes what happens first in the operation of a trolley?
- **A** Charges flow down the pole.
- **B** Charges pass through the motor.
- **C** Charges enter the shoe from the power line.
- **D** Charges return to the generator through the tracks.

Standardized Test Preparation *continued*

Passage 2 Benjamin Franklin (1706–1790) first suggested the terms *positive* and *negative* for the two types of charge. At the age of 40, Franklin was a successful printer and journalist. He saw some experiments on electricity and was so fascinated by them that he began to devote much of his time to experimenting. Franklin was the first person to realize that lightning is a huge electric discharge, or spark. He invented the first lightning rod, for which he became famous. He also flew a kite into thunderclouds—at great risk to his life—to collect charge from them. During and after the Revolutionary War, Franklin gained fame as a politician and a statesman.

_______ **1.** Which of the following happened earliest in Franklin's life?
   A He gained fame as a politician.
   B He flew a kite into thunderclouds.
   C He saw experiments on electricity.
   D He was a successful journalist.

_______ **2.** Which of the following statements is a fact according to the passage?
   F Franklin became interested in electricity in 1706.
   G There is no connection between lightning and an electric discharge.
   H Franklin became a successful journalist after he performed experiments with electricity.
   I Flying a kite into thunderclouds is dangerous.

Standardized Test Preparation *continued*

INTERPRETING GRAPHICS

Use the diagram below to answer the questions that follow.

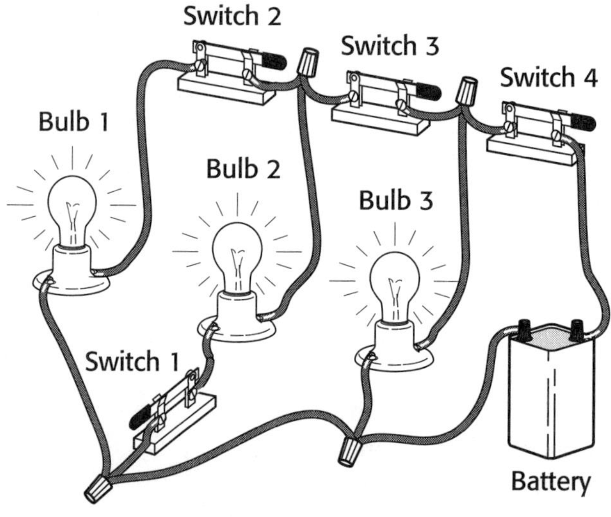

_______ **1.** Opening which switch will turn off only light bulb 2?
   A switch 1
   B switch 2
   C switch 3
   D switch 4

_______ **2.** Opening which switch will turn off exactly two light bulbs?
   F switch 1
   G switch 2
   H switch 3
   I switch 4

_______ **3.** If only switches 2 and 3 are open, which of the following will happen?
   A All three bulbs will remain lit.
   B Only bulb 1 will remain lit.
   C Only bulb 3 will remain lit.
   D All three bulbs will turn off.

_______ **4.** Which of the following statements is false?
   F Bulb 2 will be off when bulb 1 is off.
   G Bulb 3 will be on if any other bulb is on.
   H Bulbs 1 and 3 can be on when bulb 2 is off.
   I Bulb 3 can be on when bulbs 1 and 2 are off.

Standardized Test Preparation *continued*

MATH

Read each question below, and choose the best answer.

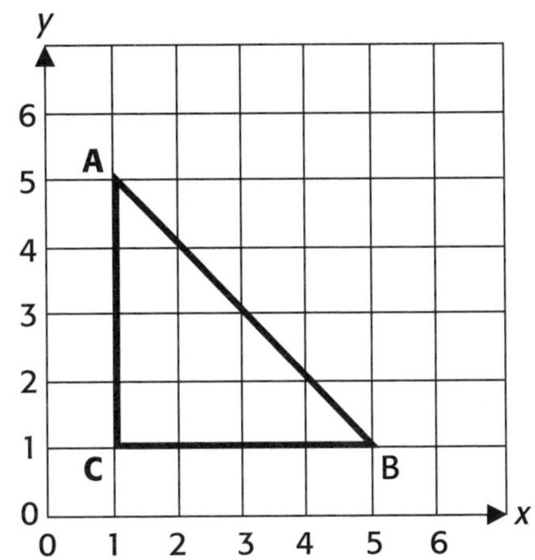

______ **1.** Look at triangle *ABC*. If you want to draw square *ADBC*, what would
the coordinates of *D* be?
  A (1, 5)
  B (3, 3)
  C (5, 5)
  D (5, 1)

______ **2.** The equation *voltage = current × resistance* is often called *Ohm's
law*. If the current in an object is 0.5 A and the voltage across the
object is 12 V, what is the resistance of the object?
  F 0.042 Ω
  G 6 Ω
  H 12.5 Ω
  I 24 Ω

______ **3.** Heather has six large dogs. In one day, each of the dogs eats 2.1 kg of
dog food. Which is the best estimate of the total number of kilograms
of food all of the dogs eat in 4 weeks?
  A less than 150 kg
  B between 150 and 225 kg
  C between 225 and 300 kg
  D more than 300 kg

Circuitry 101

There are two basic types of electric circuits. A series circuit connects all of the parts in a single loop, and a parallel circuit connects each part on a separate branch. A switch wired in series with the energy source can control the whole circuit. If you want each part of the circuit to work on its own, the loads must be wired in parallel. In this lab, you will use an ammeter to measure current and a voltmeter to measure voltage. For each circuit, you will use Ohm's law (resistance equals voltage divided by current) to determine the overall resistance.

OBJECTIVES

Build a series circuit and a parallel circuit.

Use Ohm's law to calculate the resistance of a circuit from voltage and current.

MATERIALS

- ammeter
- energy source—dry cell(s)
- light-bulb holders (3)
- light bulbs (3)
- switch
- voltmeter
- wire, insulated, 15 cm lengths with both ends stripped

SAFETY INFORMATION 

PROCEDURE

1. Build a series circuit with an energy source, a switch, and three light bulbs. Draw a diagram of your circuit. Caution: Always leave the switch open when building or changing the circuit. Close the switch only when you are testing or taking a reading.

2. Test your circuit. Do all three bulbs light up? Are all bulbs the same brightness? What happens if you carefully unscrew one light bulb? Does it make any difference which bulb you unscrew? Record your observations.

____

____

____

Circuitry 101 *continued*

---

3. Connect the ammeter between the power source and the switch. Close the switch, and record the current on your diagram. Be sure to show where you measured the current.

4. Reconnect the circuit so that the ammeter is between the first and second bulbs. Record the current, as you did in step 3.

5. Move the ammeter so that it is between the second and third bulbs, and record the current again.

6. Remove the ammeter from the circuit. Connect the voltmeter to the two ends of the power source. Record the voltage on your diagram.

7. Use the voltmeter to measure the voltage across each bulb. Record each reading.

8. Take apart your series circuit. Reassemble the same items so that the bulbs are wired in parallel. (Note: The switch must remain in series with the power source to be able to control the whole circuit.) Draw a diagram of your circuit.

9. Test your circuit, and record your observations, as you did in step 2.

____

____

____

10. Connect the ammeter between the power source and the switch. Record the current.

11. Reconnect the circuit so that the ammeter is right next to one of the three bulbs. Record the current.

12. Repeat step 11 for the two remaining bulbs.

13. Remove the ammeter from your circuit. Connect the voltmeter to the two ends of the power source. Record the voltage.

14. Measure and record the voltage across each light bulb.

Circuitry 101 *continued*

ANALYZE THE RESULTS

1. Recognizing Patterns Was the current the same at all places in the series circuit? Was it the same everywhere in the parallel circuit?

____

____

2. Analyzing Data For each circuit, compare the voltage across each light bulb with the voltage at the power source.

____

____

____

3. Identifying Patterns What is the relationship between the voltage at the power source and the voltages at the light bulbs in a series circuit?

____

____

4. Analyzing Data Use Ohm's law and the readings for current *(I)* and voltage *(V)* at the power source for both circuits to calculate the total resistance *(R)* in both the series and parallel circuits.

DRAW CONCLUSIONS

5. Drawing Conclusions Was the total resistance for both circuits the same? Explain your answer.

____

____

____

6. Interpreting Information Why did the bulbs differ in brightness?

____

____

____

Circuitry 101 *continued*

7. Making Predictions Based on your results, what do you think might happen if too many electrical appliances are plugged into the same series circuit? What might happen if too many electrical appliances are plugged into the same parallel circuit?

____

____

____

____

____

____

Quick Lab

Detecting Charge

MATERIALS

- aluminum foil (2 pieces)
- index card
- jar, glass, medium-sized
- objects that can be charged (balloons; rubber, plastic, and glass rods; plastic ruler; wool and silk cloth)
- paper clip
- scissors

SAFETY INFORMATION 

PROCEDURE

1. Use **scissors** to cut two strips of **aluminum foil** that are 1 cm × 4 cm each.

2. Bend a **paper clip** to make a hook. (The clip will look like an upside-down question mark.)

3. Push the end of the hook through the middle of an **index card**, and tape the hook so that it hangs straight down from the card.

4. Lay the two foil strips on top of one another, and hang them on the hook by gently pushing the hook through them.

5. Lay the card over the top of a **glass jar.**

6. Bring **various charged objects** near the top of the paper-clip hook, and observe what happens. Explain your observations.

____

____

____

Quick Lab

A Series of Circuits

MATERIALS

- battery, 6 V
- flashlight bulb, burned-out
- flashlight bulb with holder (3)
- screwdriver
- tape
- wire, copper, insulated with ends stripped (5)

SAFETY INFORMATION 

PROCEDURE

1. Connect a **6 V battery** and **two flashlight bulbs** in a series circuit. Draw a picture of your circuit.

2. Add **another flashlight bulb** in series with the other two bulbs. How does the brightness of the light bulbs change?

   ____

   ____

3. Replace one of the light bulbs with a **burned-out light bulb.** What happens to the other lights in the circuit? Why?

   ____

   ____

   ____

Quick Lab

A Parallel Lab

MATERIALS

- battery, 6 V
- flashlight bulb, burned-out
- flashlight bulb with holder (3)
- screwdriver
- tape
- wire, copper, insulated with ends stripped, long (2)
- wire, copper, insulated with ends stripped, short (6)

SAFETY INFORMATION 

PROCEDURE

1. Connect a **6 V battery** and **two flashlight bulbs** in a parallel circuit. Draw a picture of your circuit.

2. Add **another flashlight bulb** in parallel with the other two bulbs. How does the brightness of the light bulbs change?

____

____

____

3. Replace one of the light bulbs with a **burned-out light bulb.** What happens to the other lights in the circuit? Why?

____

____

____

Stop the Static Electricity!

Imagine this scenario: Some of your clothes cling together when they come out of the dryer. This annoying problem is caused by static electricity—the buildup of electric charges on an object. In this lab, you'll discover how this buildup occurs.

MATERIALS

- cloth, silk
- cloth, woolen
- packing peanut, plastic-foam
- rod, glass
- rod, rubber
- tape
- thread, 30 cm

SAFETY INFORMATION 

Using Scientific Methods

ASK A QUESTION

1. How do electric charges build up on clothes in a dryer?

FORM A HYPOTHESIS

2. Write a statement that answers the question above. Explain your reasoning.

____

TEST THE HYPOTHESIS

3. Tie a piece of thread approximately 30 cm in length to a packing peanut. Hang the peanut by the thread from the edge of a table. Tape the thread to the table.

4. Rub the rubber rod with the wool cloth for 10 to 15 s. Bring the rod near, but do not touch, the peanut. Observe the peanut, and record your observations. If nothing happens, repeat this step.

____

5. Touch the peanut with the rubber rod. Pull the rod away from the peanut, and then bring it near again. Record your observations.

____

6. Repeat steps 4 and 5 with the glass rod and silk cloth.

____

____

Stop the Static Electricity! *continued*

7. Now, rub the rubber rod with the wool cloth, and bring the rod near the peanut again. Record your observations.

ANALYZE THE RESULTS

1. What caused the peanut to act differently in steps 4 and 5?

2. Did the glass rod have the same effect on the peanut as the rubber rod did? Explain how the peanut reacted in each case.

3. Was the reaction of the peanut the same in steps 5 and 7? Explain.

DRAW CONCLUSIONS

4. Based on your results, was your hypothesis correct? Explain your answer, and write a new statement if necessary.

APPLYING YOUR DATA

Do some research to find out how a dryer sheet helps stop the buildup of electric charges in the dryer.

Potato Power

DATASHEET FOR LABBOOK

Have you ever wanted to look inside a D cell from a flashlight or an AA cell from a portable radio? All cells include the same basic components, as shown below. There is a metal "bucket," some electrolyte (a paste), and a rod of some other metal (or solid) in the middle. Even though cell construction is simple, companies that manufacture cells are always trying to make a product with the highest voltage possible from the least expensive materials. Sometimes, companies try different pastes, and sometimes they try different combinations of metals. In this lab, you will make your own cell. Using inexpensive materials, you will try to produce the highest voltage you can.

MATERIALS

- metal strips, labeled
- potato
- ruler, metric
- voltmeter

SAFETY INFORMATION 

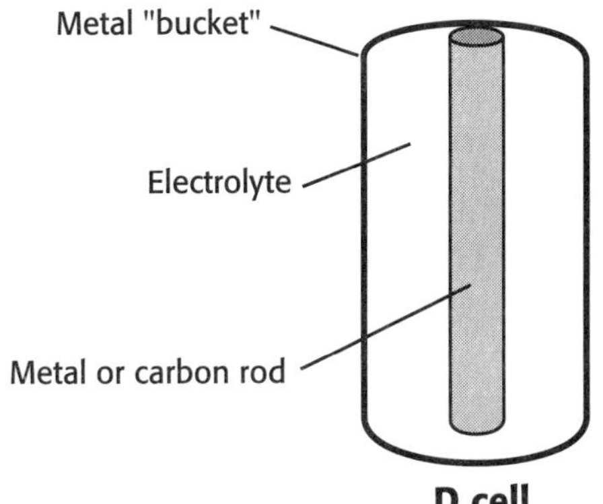

PROCEDURE

1. Choose two metal strips. Carefully push one of the strips into the potato at least 2 cm deep. Insert the second strip the same way, and measure how far apart the two strips are. (If one of your metal strips is too soft to push into the potato, push a harder strip in first, remove it, and then push the soft strip into the slit.) Record the two metals you have used and the distance between the strips. **Caution:** The strips of metal may have sharp edges.

____

____

2. Connect the voltmeter to the two strips, and record the voltage.

____

3. Move one of the strips closer to or farther from the other. Measure the new distance and voltage. Record your results.

____

| Potato Power *continued*

---

4. Repeat steps 1 through 3, using different combinations of metal strips and distances until you find the combination that produces the highest voltage.

____

____

____

ANALYZE THE RESULTS

1. What combination of metals and distance produced the highest voltage?

____

____

2. If you change only the distance but use the same metal strips, what is the effect on the voltage?

____

____

3. One of the metal strips tends to lose electrons, and the other tends to gain electrons. What do you think would happen if you used two strips of the same metal?

____

____

____

Vocabulary Activity

An Electrifying Puzzle

After you finish reading the chapter, give this crossword puzzle a try!

ACROSS

3. type of circuit in which different loads are on separate branches

4. a material in which charges cannot easily move

5. the rate at which charges pass a given point

6. The energy per unit charge is called the _____ difference.

7. The law of electric _____s states that like charges repel and opposite charges attract.

11. a device in a circuit that uses electrical energy to do work

15. converts thermal energy into electrical energy

17. consists of several cells

18. a complete, closed path through which electric charges flow

19. the opposition to the flow of electric charge

20. used to open and close a circuit

DOWN

1. a device that produces an electric current by converting chemical energy into electrical energy

2. Electric _____ is the loss of static electricity as charges move off an object.

3. the part of a solar panel that absorbs light and converts it into electrical energy

5. transfer of electrons from one object to another by direct contact

8. rearrangement of electrons on an uncharged object without direct contact with a charged object

9. A charged object exerts an electric _____ on other charged objects.

10. Electric _____ is the rate at which electrical energy does work.

12. a material in which charges can move easily

13. The buildup of electric charges on an object is called _____ electricity.

14. the difference in energy per unit charge as a charge moves between two points in the path of a circuit

16. type of circuit in which all parts are connected in a single loop

▌Vocabulary Activity *continued*

SciLinks Activity

Electric Circuits

Go to www.scilinks.org. To find links related to electric circuits, type in the keyword HSM0471. Then, use the links to answer the following questions about electric circuits.

1. What is a diode?

____

____

____

2. How does a crystal radio work?

____

____

____

3. How could you make an electric circuit with a lemon or a potato?

____

____

____

4. What will happen if you place a magnetic compass in an electric circuit?

____

____

____

Performance-Based Assessment

Teacher Notes and Answer Key

PURPOSE

Students will build simple circuits with resistors in series and in parallel. They will use these circuits to answer questions about the relationship between resistor combinations and voltage at different points in a circuit.

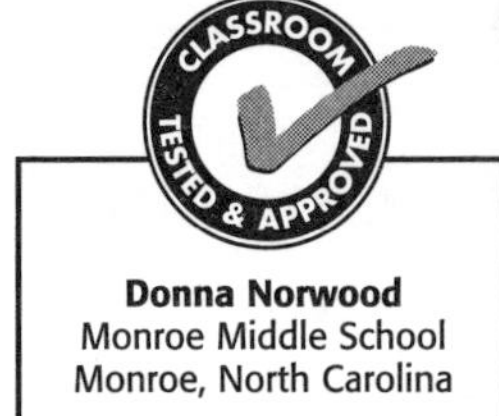

Donna Norwood
Monroe Middle School
Monroe, North Carolina

TIME REQUIRED

One 45-minute class period
Students will need 25 minutes at the activity station and 20 minutes to answer the analysis questions.

RATING

Easy ← 1　2　3　4 → Hard

Teacher Prep–2
Student Set-up–2
Concept Level–3
Clean Up–1

ADVANCE PREPARATION

Equip each activity station with the necessary materials.

SAFETY INFORMATION

Conduct this activity in an area clear of excess materials. Falling, dropped, or rolling objects are a slipping or tripping hazard and could cause injury.

TEACHING STRATEGIES

This activity works best in groups of 2–3 students. If a voltage V_0 is put across several resistors in a series, each resistor drops the voltage in proportion to that resistor's contribution to the total resistance. That is, the voltage drop across R_1 equals $V_0 \times R_1/(R_1 + R_2 + \ldots + R_n)$. Therefore, with three identical resistors R, the voltage drops by $V_0 \times R/(3R) = (1/3) V_0$ across each resistor. Also, if n identical resistors R are placed in parallel, they act like a single resistor with resistance R/n.

Performance-Based Assessment *continued*

Evaluation Strategies

Use the following rubric to help evaluate student performance.

Rubric for Assessment

Possible points	Appropriate use of materials and equipment (20 points possible)
20–14	Successfully completes activity; safe and careful use of materials and equipment; attention to detail; superior lab skills
13–7	Generally completes activity; adequate use of materials and equipment; moderate attention to detail; sound knowledge of lab techniques
6–1	Does not complete activity; sloppy use of materials and equipment; no attention to detail; apparent lack of skill
	Quality and clarity of measurements (40 points possible)
40–30	Correct measurements; stated clearly and accurately
29–20	Largely correct measurements; stated less clearly
19–10	A few measurements missing; some confusion in presentation
9–1	Measurements largely incomplete
	Explanation of observations (40 points possible)
40–30	Clear, detailed explanations; superior insight into how resistors and circuits work
29–20	Less-clear, less-detailed explanations; good insight into how resistors and circuits work
19–10	Unclear explanations; flawed insight into how resistors and circuits work
9–1	Poor explanations; poor grasp of how resistors and circuits work

Name _____________________________ Class _____________ Date _____________

SKILLS PRACTICE

Performance-Based Assessment

OBJECTIVE

In this activity, you will connect resistors in series and in parallel and note how each arrangement affects the voltage readings in simple electric circuits.

KNOW THE SCORE!

As you work through the activity, keep in mind that you will be earning a grade for the following:

- how well you work with the materials and choose the correct scenario (20%)
- how well you explain your observations (40%).
- whether you construct an operational model (40%)

Using Scientific Methods

ASK A QUESTION

How does a parallel circuit differ from a series circuit?

MATERIALS AND EQUIPMENT

- Broomstick or wooden dowel
- 6 V battery
- 2 wires with alligator clips
- 1,000 ohm resistors (3)
- voltmeter

SAFETY INFORMATION

Keeping your work area clean helps maintain a safe work environment.

FORM A HYPOTHESIS

You will be attaching resistors to terminals of the battery. How do you think this will affect the voltage?

TEST THE HYPOTHESIS

1. Attach three resistors between the terminals of the battery as shown in diagram A at left. The resistors are now *in series*: all of the current in the circuit flows through each resistor.

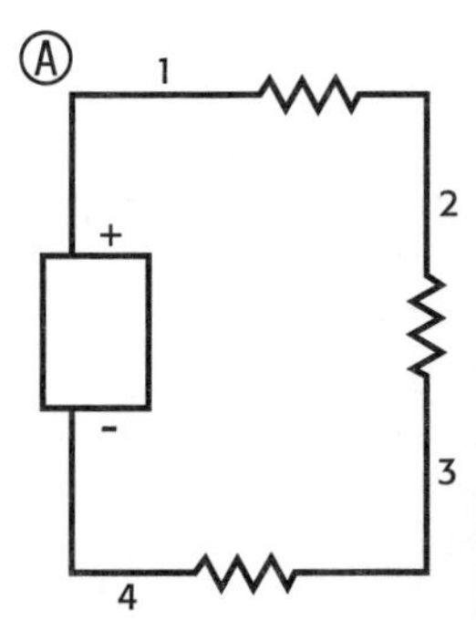

Name _____________________________ Class _______________ Date _______________

Performance-Based Assessment *continued*

2. Attach the black probe of the voltmeter to the negative terminal of the battery. With the red probe, measure the voltage in the circuit at each of the four points shown in the diagram. Record your data below.

Sample answer: The voltage is 6 V at point 1, 4.5 V at point 2, 2 V at point 3,

and 0 V at point 4.

3. Disassemble the circuit. Now, attach three resistors between the terminals of the battery as shown in diagram B. The two resistors between points 2 and 3 are now *in parallel*: half the circuit current passes through each of them. The entire current still passes through the resistor between points 1 and 2.

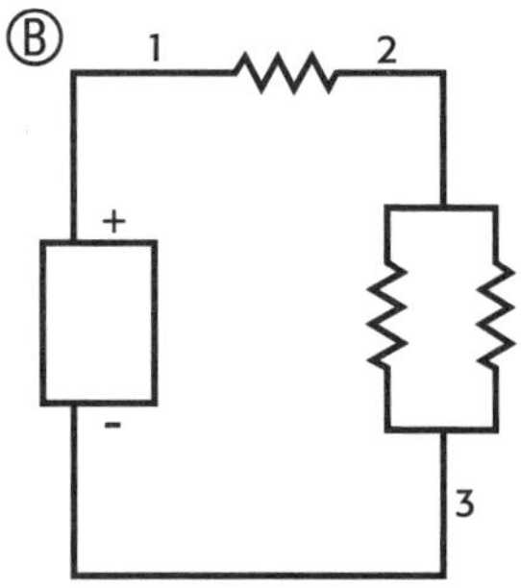

4. Attach the black probe of the voltmeter to the negative terminal of the battery. With the red probe, measure the voltage in the circuit at each of the three points shown in the diagram. Record your data below.

Sample answer: 6 V at point 1,2 V at point 2,0 V at point 3

ANALYZE THE RESULTS

5. Compare the voltage drop between points 1 and 2 with the voltage drop between points 2 and 4 when the resistors are in series. Explain what is happening.

The voltage drops twice as much between points 2 and 4 as it does between

points 1 and 2 because the current has to go through twice as much

resistance. This suggests that when resistors are in series, the effect is

additive—the total resistance is the sum of the individual resistances.

6. Compare the drop in voltage between points 1 and 2 with the drop between points 2 and 3 when the resistors are in parallel.

The voltage drops twice as much between points 1 and 2 as between 2 and 3.

Name _________________________________ Class _______________ Date _____________

Performance-Based Assessment *continued*

7. What is the overall resistance between points 2 and 3 when the resistors are in parallel?

Sample answer: The resistance between points 1 and 2 is twice as great as between 2 and 3, which means that the overall resistance between points 2 and 3 must be 500 ohms.

DRAW CONCLUSIONS

8. What differences do you see in your results? Why do you think these differences occurred?

The voltage drops as the amount of resistance increases. The resistance slows down the current which affects the voltage.

DATASHEET FOR CHAPTER LAB

Circuitry 101

Teacher Notes and Answer Key

TIME REQUIRED
Two 45-minute class periods

LAB RATINGS

Easy ◄——1——2——3——4——► Hard

Teacher Prep–2
Student Set-Up–2
Concept Level–3
Clean Up–1

MATERIALS

Materials listed are for each group of 3–5 students. Use a series of three 1.5 V cells or a 6 V battery as a power source. If you use a series of 1.5 V cells, create a single battery by taping the cells together (positive to negative). Use only regular dry cell or alkaline batteries. Do not use nicad or lithium batteries.

SAFETY CAUTION

Remind students to review all safety cautions and icons before beginning this lab activity. Caution students that the meters can be damaged if connected improperly.

PROCEDURE NOTES

Tell or show students how to build a circuit. Demonstrate the proper way to use an ammeter and a voltmeter. You may wish to perform this lab as a demonstration if materials are limited.

Name _______________________________ Class _______________ Date _______________

Skills Practice

DATASHEET FOR CHAPTER LAB

Circuitry 101

There are two basic types of electric circuits. A series circuit connects all of the parts in a single loop, and a parallel circuit connects each part on a separate branch. A switch wired in series with the energy source can control the whole circuit. If you want each part of the circuit to work on its own, the loads must be wired in parallel. In this lab, you will use an ammeter to measure current and a voltmeter to measure voltage. For each circuit, you will use Ohm's law (resistance equals voltage divided by current) to determine the overall resistance.

OBJECTIVES

Build a series circuit and a parallel circuit.

Use Ohm's law to calculate the resistance of a circuit from voltage and current.

MATERIALS

- ammeter
- energy source—dry cell(s)
- light-bulb holders (3)
- light bulbs (3)

- switch
- voltmeter
- wire, insulated, 15 cm lengths with both ends stripped

SAFETY INFORMATION 

PROCEDURE

1. Build a series circuit with an energy source, a switch, and three light bulbs. Draw a diagram of your circuit. Caution: Always leave the switch open when building or changing the circuit. Close the switch only when you are testing or taking a reading.

2. Test your circuit. Do all three bulbs light up? Are all bulbs the same brightness? What happens if you carefully unscrew one light bulb? Does it make any difference which bulb you unscrew? Record your observations.

____

____

____

Name _________________________________ Class _______________ Date _______________

Circuitry 101 *continued*

3. Connect the ammeter between the power source and the switch. Close the switch, and record the current on your diagram. Be sure to show where you measured the current.

4. Reconnect the circuit so that the ammeter is between the first and second bulbs. Record the current, as you did in step 3.

5. Move the ammeter so that it is between the second and third bulbs, and record the current again.

6. Remove the ammeter from the circuit. Connect the voltmeter to the two ends of the power source. Record the voltage on your diagram.

7. Use the voltmeter to measure the voltage across each bulb. Record each reading.

8. Take apart your series circuit. Reassemble the same items so that the bulbs are wired in parallel. (Note: The switch must remain in series with the power source to be able to control the whole circuit.) Draw a diagram of your circuit.

9. Test your circuit, and record your observations, as you did in step 2.

____

____

____

10. Connect the ammeter between the power source and the switch. Record the current.

11. Reconnect the circuit so that the ammeter is right next to one of the three bulbs. Record the current.

12. Repeat step 11 for the two remaining bulbs.

13. Remove the ammeter from your circuit. Connect the voltmeter to the two ends of the power source. Record the voltage.

14. Measure and record the voltage across each light bulb.

Name _______________________________ Class _______________ Date ______________

Circuitry 101 *continued*

ANALYZE THE RESULTS

1. **Recognizing Patterns** Was the current the same at all places in the series circuit? Was it the same everywhere in the parallel circuit?

   The current was the same at all points in the series circuit. The current was

   not the same at all points in the parallel circuit.

2. **Analyzing Data** For each circuit, compare the voltage across each light bulb with the voltage at the power source.

   The voltage across each light bulb in the series circuit was about the same

   and less than the voltage at the power source. The voltage across each light

   bulb in the parallel circuit was the same as the voltage at the power source.

3. **Identifying Patterns** What is the relationship between the voltage at the power source and the voltages at the light bulbs in a series circuit?

   The voltage at the power source is the sum of the voltages at the light bulbs

   in a series circuit.

4. **Analyzing Data** Use Ohm's law and the readings for current *(I)* and voltage *(V)* at the power source for both circuits to calculate the total resistance *(R)* in both the series and parallel circuits.

   Answers will depend on the bulbs used. Check for correct calculations.

DRAW CONCLUSIONS

5. **Drawing Conclusions** Was the total resistance for both circuits the same? Explain your answer.

   No. The series circuit had a much higher resistance than the parallel circuit.

   The resistance is less in a parallel circuit because the electric current has

   more than one path to follow.

6. **Interpreting Information** Why did the bulbs differ in brightness?

   The parallel circuit had brighter bulbs because the voltage across each bulb

   was greater than in the series circuit.

Name _______________________________ Class _______________ Date _______________

Circuitry 101 *continued*

7. Making Predictions Based on your results, what do you think might happen if too many electrical appliances are plugged into the same series circuit? What might happen if too many electrical appliances are plugged into the same parallel circuit?

Too many appliances on the same series circuit would cause each appliance to get very little voltage, so they may not receive enough electric current to work. Too many appliances on the same parallel circuit could cause too much electric current in the circuit (because all loads in a parallel circuit draw the same voltage), leading to an overloaded circuit.

Name _______________________________ Class ______________ Date ___________

Detecting Charge

DATASHEET FOR QUICK LAB

MATERIALS

- aluminum foil (2 pieces)
- index card
- jar, glass, medium-sized
- objects that can be charged (balloons; rubber, plastic, and glass rods; plastic ruler; wool and silk cloth)
- paper clip
- scissors

SAFETY INFORMATION 

PROCEDURE

1. Use **scissors** to cut two strips of **aluminum foil** that are 1 cm × 4 cm each.
2. Bend a **paper clip** to make a hook. (The clip will look like an upside-down question mark.)
3. Push the end of the hook through the middle of an **index card**, and tape the hook so that it hangs straight down from the card.
4. Lay the two foil strips on top of one another, and hang them on the hook by gently pushing the hook through them.
5. Lay the card over the top of a **glass jar.**
6. Bring **various charged objects** near the top of the paper-clip hook, and observe what happens. Explain your observations.

   <u>Students should observe that the two leaves of foil in the jar spread apart</u>

   <u>when a charged object is brought near or touched to the paper clip.</u>

Name _______________________________ Class _______________ Date _______________

Quick Lab
A Series of Circuits

MATERIALS

- battery, 6 V
- flashlight bulb, burned-out
- flashlight bulb with holder (3)
- screwdriver
- tape
- wire, copper, insulated with ends stripped (5)

SAFETY INFORMATION 

PROCEDURE

1. Connect a **6 V battery** and **two flashlight bulbs** in a series circuit. Draw a picture of your circuit.

The drawing of the circuit should show the light bulbs and the 6 V battery connected in a loop.

2. Add **another flashlight bulb** in series with the other two bulbs. How does the brightness of the light bulbs change?

The light bulbs do not glow as brightly as they did when only two bulbs were

attached.

3. Replace one of the light bulbs with a **burned-out light bulb.** What happens to the other lights in the circuit? Why?

The other light bulbs in the circuit do not glow. The burned-out bulb causes

a break in the circuit, so no charges can flow.

Name _______________________________ Class _______________ Date _____________

DATASHEET FOR QUICK LAB

A Parallel Lab

MATERIALS

- battery, 6 V
- flashlight bulb, burned-out
- flashlight bulb with holder (3)
- screwdriver
- tape
- wire, copper, insulated with ends stripped, long (2)
- wire, copper, insulated with ends stripped, short (6)

SAFETY INFORMATION 

PROCEDURE

1. Connect a **6 V battery** and **two flashlight bulbs** in a parallel circuit. Draw a picture of your circuit.

   The drawing of the circuit should show the light bulbs on different branches.

2. Add **another flashlight bulb** in parallel with the other two bulbs. How does the brightness of the light bulbs change?

   The light bulbs glow with the same brightness as they did when only two

   bulbs were attached.

3. Replace one of the light bulbs with a **burned-out light bulb.** What happens to the other lights in the circuit? Why?

   The other light bulbs in the circuit still glow because charges can flow

   through the other branches.

Skills Practice Lab

DATASHEET FOR LABBOOK

Stop the Static Electricity!

Teacher Notes and Answer Key

TIME REQUIRED

One 45-minute class period

LAB RATINGS

Easy 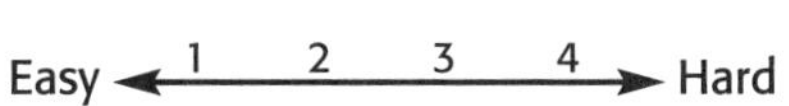 Hard

Teacher Prep–1
Student Set-Up–1
Concept Level–2
Clean Up–1

PROCEDURE NOTES

This lab works best on cold, dry days. If you live in an area with high humidity, use a dehumidifier, or substitute a piece of PVC pipe for the rubber rod.

Name _________________________________ Class _____________ Date __________

DATASHEET FOR LABBOOK

Stop the Static Electricity!

Imagine this scenario: Some of your clothes cling together when they come out of the dryer. This annoying problem is caused by static electricity—the buildup of electric charges on an object. In this lab, you'll discover how this buildup occurs.

MATERIALS

- cloth, silk
- cloth, woolen
- packing peanut, plastic-foam
- rod, glass
- rod, rubber
- tape
- thread, 30 cm

SAFETY INFORMATION 

Using Scientific Methods

ASK A QUESTION

1. How do electric charges build up on clothes in a dryer?

FORM A HYPOTHESIS

2. Write a statement that answers the question above. Explain your reasoning.

____

TEST THE HYPOTHESIS

3. Tie a piece of thread approximately 30 cm in length to a packing peanut. Hang the peanut by the thread from the edge of a table. Tape the thread to the table.

4. Rub the rubber rod with the wool cloth for 10 to 15 s. Bring the rod near, but do not touch, the peanut. Observe the peanut, and record your observations. If nothing happens, repeat this step.

____

5. Touch the peanut with the rubber rod. Pull the rod away from the peanut, and then bring it near again. Record your observations.

____

6. Repeat steps 4 and 5 with the glass rod and silk cloth.

____

____

Name _______________________________ Class _______________ Date _____________

Stop the Static Electricity! *continued*

7. Now, rub the rubber rod with the wool cloth, and bring the rod near the peanut again. Record your observations.

____

ANALYZE THE RESULTS

1. What caused the peanut to act differently in steps 4 and 5?

Step 4—the rod induced an opposite charge in the peanut; step 5—the charge

was transferred by touching.

____

2. Did the glass rod have the same effect on the peanut as the rubber rod did? Explain how the peanut reacted in each case.

yes; the peanut was initially attracted and then repelled by the rods.

____

3. Was the reaction of the peanut the same in steps 5 and 7? Explain.

The peanut was repelled by the rod in step 5. It was attracted to the rod in

step 7. The peanut's charge must have changed due to the charge on the glass

rod.

____

DRAW CONCLUSIONS

4. Based on your results, was your hypothesis correct? Explain your answer, and write a new statement if necessary.

New answers should include that charges build up through friction as clothes

rub together.

____

APPLYING YOUR DATA

Do some research to find out how a dryer sheet helps stop the buildup of electric charges in the dryer.

Dryer sheets leave a layer of soap on clothes, and this soap attracts moisture

from the air. This process allows built-up charges to dissipate.

____

Model Making Lab)

DATASHEET FOR LABBOOK

Potato Power

Teacher Notes and Answer Key

TIME REQUIRED

One 45-minute class period

Susan Gorman
Northridge Middle School
North Richland Hills, Texas

LAB RATINGS

Easy 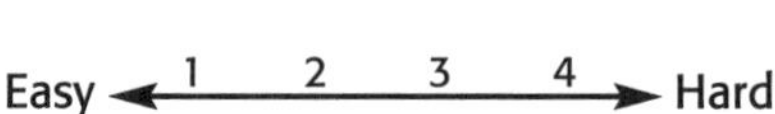 Hard

Teacher Prep–1
Student Set-Up–1
Concept Level–1
Clean Up–1

MATERIALS

Provide each group of 2–3 students with at least 3 metal strips (0.5 cm × 2 cm) of dissimilar metals. Metal strips that work well are zinc, copper, aluminum, and iron. If metal strips are difficult to obtain, copper wires, zinc screws, and other pieces of iron and aluminum may be used. A voltmeter with a range of 0–5 V works best. Students may also need 2 wires to connect the meter to the metal strips.

SAFETY CAUTION

Caution students that the metal strips may have sharp edges. Students should wash their hands thoroughly after handling the potatoes.

Name _________________________ Class ______________ Date __________

DATASHEET FOR LABBOOK

Potato Power

Have you ever wanted to look inside a D cell from a flashlight or an AA cell from a portable radio? All cells include the same basic components, as shown below. There is a metal "bucket," some electrolyte (a paste), and a rod of some other metal (or solid) in the middle. Even though cell construction is simple, companies that manufacture cells are always trying to make a product with the highest voltage possible from the least expensive materials. Sometimes, companies try different pastes, and sometimes they try different combinations of metals. In this lab, you will make your own cell. Using inexpensive materials, you will try to produce the highest voltage you can.

MATERIALS

- metal strips, labeled
- potato
- ruler, metric
- voltmeter

SAFETY INFORMATION

PROCEDURE

1. Choose two metal strips. Carefully push one of the strips into the potato at least 2 cm deep. Insert the second strip the same way, and measure how far apart the two strips are. (If one of your metal strips is too soft to push into the potato, push a harder strip in first, remove it, and then push the soft strip into the slit.) Record the two metals you have used and the distance between the strips. **Caution:** The strips of metal may have sharp edges.

____

____

2. Connect the voltmeter to the two strips, and record the voltage.

____

3. Move one of the strips closer to or farther from the other. Measure the new distance and voltage. Record your results.

____

Name _______________________________ Class _______________ Date ____________

Potato Power *continued*

4. Repeat steps 1 through 3, using different combinations of metal strips and distances until you find the combination that produces the highest voltage.

____

____

____

____

ANALYZE THE RESULTS

1. What combination of metals and distance produced the highest voltage?

Answers will depend on the metals used.

____

2. If you change only the distance but use the same metal strips, what is the effect on the voltage?

As distance increases, voltage decreases.

____

3. One of the metal strips tends to lose electrons, and the other tends to gain electrons. What do you think would happen if you used two strips of the same metal?

If you used two strips of the same metal, there would be no voltage because

each strip has the same tendency to gain or lose electrons.

____

Answer Key

Directed Reading A

SECTION: ELECTRIC CHARGE AND STATIC ELECTRICITY

1. B
2. C
3. positive charge, negative charge, or no charge
4. The law states that like charges repel and opposite charges attract.
5. positively
6. negatively
7. electric force
8. the amount of each charge and the distance between the charges
9. electric field
10. They either attract or repel one another.
11. There are equal numbers of positively charged protons and negatively charged electrons so their charges cancel each other out.
12. It becomes positively charged.
13. It becomes negatively charged.
14. C
15. B
16. A
17. created, destroyed.
18. electroscope
19. You cannot tell whether the object's charge is positive or negative.
20. A
21. B
22. C
23. B
24. Metals are good conductors because their electrons are free to move.
25. Insulators do not conduct charges well because their electrons cannot flow freely and are tightly held in the atoms.
26. Static electricity is the electric charge at rest on an object.
27. electric discharge
28. a flash of light, a shock, or a crackling noise
29. In the cloud, water, ice, and air move and build up negative charges often at the bottom of the cloud. Positive charges build on the top of the cloud. This causes a rapid electric discharge called lightning.

30. Lightning will strike the highest point in a charged area and on a beach you may be the highest point.
31. The lightning rod is the tallest point on the building so the lightning strikes it. The rod has a wire that is grounded and provides a path for the electricity to the Earth and not through the building.

SECTION: ELECTRIC CURRENT AND ELECTRICAL ENERGY

1. electrical energy
2. C
3. A
4. B
5. B
6. electric field
7. alternating
8. direct
9. Batteries use DC and outlets use AC.
10. C
11. C
12. B
13. A
14. D
15. ohms
16. current
17. the object's material, thickness, length, and temperature
18. low
19. The high resistance of the filament causes the light bulb to heat up and give off light.
20. Materials with a resistance of 0 ohms.
21. A
22. D
23. E
24. C
25. B
26. chemical energy, into electrical energy
27. In a wet cell the electrolytes are liquid. In a dry cell, the electrolytes are solid or pastelike.
28. car engines, furnaces, ovens
29. As light shines on the photocell, electrons gain enough energy to move between atoms. The electrons then move through wire to provide electrical energy.

SECTION: ELECTRICAL CALCULATIONS

1. C
2. A
3. He measured the current that resulted from different voltages applied to a piece of metal wire.
4. B
5. B
6. B
7. B
8. watt, kilowatt
9. The bulb burns brighter because more electrical energy is converted into light energy.
10. 1,000
11. power, time
12. $E = P \times t$
   (*electrical energy = power × time*)
13. kWh
14. meter
15. Answers will vary. Sample answer: turning off the lights when you leave a room, replacing appliances with high power ratings with those with lower ratings
16. Answers will vary. Sample answer: Lighting could make up as much as 25 percent of a home's energy consumption.
17. Answers will vary. Sample answer: Fluorescent light bulbs use one-quarter to one-third the amount of energy as incandescent bulbs, and last up to ten times longer. Incandescent bulbs are inefficient because they produce mostly thermal energy, not light energy.
18. The ENERGY STAR® program, supported by the United States Environmental Protection Agency and the Department of Energy, promotes the efficient use of energy by providing energy efficiency guidelines for household electrical devices.

SECTION: ELECTRIC CIRCUITS

1. circuit
2. B
3. B
4. D
5. battery, photocell, thermocouple, electric generator
6. switch
7. If it closed, the charges will flow through the circuit.
8. If it is open, the circuit is broken and the charges stop flowing through the circuit.
9. loads
10. series circuit
11. The resistance of the whole circuit goes up and the current drops.
12. The charges will stop flowing.
13. If a refrigerator and a lamp were wired on the same series circuit, the refrigerator would only work if the lamp was on.
14. parallel circuit
15. voltage
16. The charges can still run through the other branches.
17. Each electrical outlet is on its own branch and has its own switch.
18. C
19. Answers will vary. Sample answer: broken wires or water
20. If the current is too hot, the metal strip melts causing a break in the circuit.
21. circuit breaker
22. If the current on one side is different from the current in the other side, the GFCI opens the circuit and the charges stop flowing.
23. Answers will vary. Sample answer: Make sure insulation on cords is not worn. Do not overload circuits by plugging in too many electrical devices.

Directed Reading B

SECTION: ELECTRIC CHARGE AND STATIC ELECTRICITY

1. B
2. C
3. B
4. C
5. C
6. B
7. A
8. D
9. A
10. B
11. C
12. D

13. B
14. A
15. A
16. B
17. C
18. B
19. lightning
20. lightning rod
21. grounded

SECTION: ELECTRIC CURRENT AND ELECTRICAL ENERGY

1. D
2. C
3. A
4. B
5. electron
6. alternating
7. switch
8. direct
9. A
10. C
11. B
12. D
13. A
14. C
15. B
16. D
17. C
18. A
19. B
20. A
21. C
22. B
23. D
24. A
25. C

SECTION: ELECTRICAL CALCULATIONS

1. A
2. C
3. A
4. C
5. B
6. A
7. D
8. B
9. B
10. C
11. A
12. B
13. D
14. B
15. C
16. A

SECTION: ELECTRIC CIRCUITS

1. A
2. B
3. B
4. D
5. A
6. C
7. D
8. A
9. B
10. B
11. A
12. B
13. C
14. A
15. B
16. parallel
17. loads
18. series
19. voltage
20. switch
21. C
22. B
23. B
24. B
25. A
26. C

Vocabulary and Section Summary

SECTION: ELECTRIC CHARGE AND STATIC ELECTRICITY

1. law of electric charges: the law that states that like charges repel and opposite charges attract
2. electric force: the force of attraction or repulsion on a charged particle that is due to an electric field
3. electric field: the space around a charged object in which another charged object experiences an electric force
4. electrical conductor: a material in which charges can move freely
5. electrical insulator: a material in which charges cannot freely move
6. static electricity: electric charge at rest; generally produced by friction or induction
7. electric discharge: the release of electricity stored in a source

SECTION: ELECTRIC CURRENT AND ELECTRICAL ENERGY

1. electric current: the rate at which charges pass through a given point; measured in amperes

2. voltage: the potential difference between two points; measured in volts
3. resistance: in physical science, the opposition presented to the current by a material or device
4. cell: in electricity, a device that produces an electric current by converting chemical or radiant energy into electrical energy
5. thermocouple: a device that converts thermal energy into electrical energy
6. photocell: a device that converts light energy into electrical energy

SECTION: ELECTRICAL CALCULATIONS

1. electric power: the rate at which electrical energy is converted into other forms of energy

SECTION: ELECTRIC CIRCUITS

1. series circuit: a circuit in which the parts are joined one after another such that the current in each part is the same
2. parallel circuit: a circuit in which the parts are joined in branches such that the potential difference across each part is the same

Section Review

SECTION: ELECTRIC CHARGE AND STATIC ELECTRICITY

1. Static electricity is the buildup of electric charges on an object. Electric discharge happens when the electric charges move off an object.
2. Electric force is the push or pull on a charged object that is a result of an electric field.
3. Electric charges move freely in an electrical conductor. Electric charges do not move freely in an electrical insulator.
4. B
5. Charging by friction happens when charges are transferred from one object to another as they rub together. Charging by conduction happens when charges move between a charged object and a neutral object. Charging by induction happens when the charges in a metal object are rearranged as a result of a nearby charged object.
6. The law of electric charges says that two objects that are positively charged will repel one another.
7. Examples of static electricity include static cling and a charged balloon.
8. Examples of electric discharge include lightning and the spark when you reach out for a doorknob after walking across a carpet.
9. You cannot tell whether the charge is positive or negative. The leaves of the electroscope will spread apart no matter which charge is present.
10. The metal rod is a conductor, so electrons can easily move down the rod to the metal leaves or up the rod from the metal leaves and confirm any charge present. The rubber stopper is an insulator. Electrons will not move through rubber, so the electroscope will not perform as intended.
11. The balloons have the same charge. Each balloon is pushing the other away.
12. The image would look the same. The balloons have the same charge. Giving each balloon the opposite charge to the charge it has now would mean that the balloons would still have the same charge and would repel one another.

SECTION: ELECTRIC CURRENT AND ELECTRICAL ENERGY

1. electric current
2. resistance
3. voltage
4. cell
5. D
6. A cell is made of an electrolyte and two electrodes. Chemical reactions in the electrolyte leave extra electrons on one electrode and strip them from the other. If the charged electrodes are connected with a wire, electric charges will flow between them.
7. Charges in an alternating current reverse direction many times a second. Charges in direct current move in only one direction.
8. Under equal conditions, the current from a 12 V car battery is greater than the current from a 1.5 V flashlight cell. (Students will learn in the next section that resistance might change this comparison.)

9. Increasing the resistance decreases the current.
10. Sample answer: The prefix *photo-* refers to light, as in the terms *photosynthesis* or *photograph*. So, a photocell must take in light energy. The prefix *thermo-* refers to heat, as in the term *thermometer*. So, a thermocouple must take in thermal energy.
11. Some solar calculators contain batteries as a backup power source, for when there is not enough light.
12. A short, thick copper wire at a low temperature would have a lower resistance than a long, thin iron wire at a high temperature.
13. Wire A should have the lower resistance because it is shorter than wire B. The resistance of a wire increases as the length of the wire increases.

SECTION: ELECTRICAL CALCULATIONS

1. Sample answer: Electric power is how fast electrical energy is changed into other forms of energy.
2. D
3. Circuit B
4. C
5. Sample answer: Two ways to save energy at home are to use a fan instead of an air conditioner and to turn off lights when they are not in use.
6. Sample answer: An energy-saving device that is available today is a household fluorescent light bulb.
7. $V = I \times R = 3\,A \times 9 = 27\,V$
8. $E = P \times t = 0.040kW \times 12h = 0.48\ kWh$
9. Increasing the voltage applied to a wire will increase the current. Reducing the resistance of the wire would also increase the current.
10. Substituting $P = V \times I$ into $E = P \times t$ results in the equation $E = V \times I \times t$. Substituting $V = I \times R$ into the new equation for electrical energy results in the equation $E = I \times R \times I \times t$. This can be arranged to $E = I^2 \times R \times t$.

SECTION: ELECTRIC CIRCUITS

1. Sample answer: A series circuit is a circuit in which the loads are joined one after another like the cars of a train. A parallel circuit is a circuit in which the loads are joined so that each load is on its own branch, like the wooden railroad ties under the train tracks.
2. D
3. The energy source provides electrical energy. The load is any device that uses the electrical energy to do work. Wires connect the energy source to the load.
4. Fuses and circuit breakers create breaks in a circuit when the current gets too high, preventing charges from flowing. "Breaking" the circuit prevents overheating and fires.
5. One possible reason is that there is a short circuit in the heater. Another possible reason is that the heater, along with any other devices attached to the circuit, overloads the circuit and trips the breaker.
6. No, a fuse in parallel with the device it is supposed to protect will not work successfully. If the fuse should blow, it would open only the branch that the fuse is on. The device will continue running because it is on a separate branch.
7. The circuit in A is a series circuit. The circuit in B is a parallel circuit.

Chapter Review

1. electrical conductor
2. electric discharge
3. cell
4. Resistance
5. Electric power
6. series circuit
7. D
8. B
9. A
10. C
11. A
12. A
13. B
14. When a switch is open, it creates a break in the circuit. Charges cannot flow in the circuit. When a switch is closed, it closes the gap in the circuit. The circuit is complete, so charges can flow in it.

15. One factor is the amount of the electric charge. The greater the charge is, the greater the force. The other factor is the distance between the charges. The closer the charges are to each other, the greater the force is.
16. The charges in direct current flow in one direction. In alternating current, the charges continually switch from flowing in one direction to flowing in the reverse direction.
17. The power rating of an electrical device describes the rate at which the device uses electrical energy. A clothes dryer is an example of an electrical device that has a high power rating.
18. $V = I \times R = 6 \text{ A} \times 3 \ \Omega = 18 \text{ V}$
19. $I = V/R = 60 \text{ V}/15 \ \Omega = 4 \text{ A}$
20. $R = V/I = 40 \text{ V}/5 \text{ A} = 8 \ \Omega$
21. $I = P/V = 150 \text{ W}/120 \text{ V} = 1.25 \text{ A}$
22. An answer to this exercise can be found at the end of this Teacher Edition.
23. The electrician must have wired the fish-tank bubbler in series with the lights and the computer in series with the overhead projector.
24. You would push the strip of copper and the strip of silver into the apple. The apple is the electrolyte, and the metal strips are the electrodes. Students may identify the cell as a dry cell because the apple is a solid or as a wet cell because the apple juice conducts the electric current.
25. When the pipe is rubbed with a piece of wool, the pipe is charged by friction. When the charged pipe is held close to the can, the charges in the can are rearranged and the can is charged by induction. The side of the can closest to the pipe has the opposite charge that the pipe has and the can is attracted to the pipe.
26. conductors: tap water in glass, wrench, metal part of scissors, liquid soap in plastic bottle; insulators: basketball, glass, plastic bottle, plastic scissors handles, wooden table

Reinforcement

CHARGE!
1. electric discharge
2. conduction
3. conduction
4. induction or friction
5. friction
6. induction

ELECTRIC CIRCUITS

1. A	6. series
2. D	7. wet
3. B	8. 100 W
4. E	9. 12 V
5. C	

10. The current is 8.33 amperes.
11. The resistance is 1.44 ohms.
12. dry

Critical Thinking

1. Answers will vary. Sample answer: Bald eagles and other large birds are much more likely to live in this type of habitat in developed areas.
2. Birds have a small resistance to current, according to Ohm's law
3. The metal crossbar goes to the ground, so the voltage of the metal crossbar must be zero.
4. When an eagle puts one talon on the electric wire and the other talon on the metal crossbar, a potential difference is created between the eagle's talons, allowing a current to flow. Because an eagle has a low resistance, a very large current can move instantly through the eagle and electrocute it.
5. The voltage is the same at each foot. The potential difference is calculated as follows: $13{,}000 \ V - 13{,}000 \ V = 0 \ V$. There is no potential difference between the bird's feet.
6. When a bird is perched on one power line without touching any other lines or the metal crossbar, there is no potential difference between the feet of the bird. Because there is no potential difference, there is no current flowing through the bird, according to Ohm's law.

Section Quizzes

SECTION: ELECTRIC CHARGE AND STATIC ELECTRICITY

1. B
2. C
3. A
4. D
5. C
6. A
7. B

SECTION: ELECTRIC CURRENT AND ELECTRICAL ENERGY

1. C
2. B
3. D
4. E
5. D
6. B
7. C
8. F
9. G
10. A

SECTION: ELECTRICAL CALCULATIONS

1. B
2. A
3. D
4. C
5. A
6. D

SECTION: ELECTRIC CIRCUITS

1. C
2. A
3. A
4. B
5. D
6. A
7. C

Chapter Test A

1. C
2. A
3. B
4. D
5. B
6. A
7. D
8. A
9. J
10. D
11. C
12. B
13. H
14. I
15. F
16. G
17. E
18. A
19. C
20. B
21. B
22. D
23. C
24. B
25. A

Chapter Test B

1. electrical insulators
2. electric current
3. series circuit
4. electric power
5. static electricity
6. C
7. B
8. C
9. A
10. B
11. Ohm's law gives the relationship between current, voltage, and resistance. It is summarized by the formula $I = V/R$; the current (I) equals voltage (V) divided by resistance (R).
12. Answers will vary. Sample answer: Incandescent light bulbs are inexpensive, but are inefficient because they produce mostly thermal energy, not light energy. Fluorescent bulbs use only one-quarter to one-third the amount of energy as incandescent bulbs, and can last up to ten times longer.
13. The circuit breaker is a switch that automatically open when the current is too high. A strip of metal in the breaker warms up, bends, and opens the switch, which opens the circuit. The charges then stop flowing.
14. The burned out bulb is a break in the circuit, so the circuit is no longer a closed path through which electric charges may flow.
15. By Ohm's law, $R = V/I$. The resistance of the circuit is 3.0 V/1.5 A = 2 Ω.
16. The metal strip in the fuse melted because too much current was flowing through the circuit. If the fuse had not blow, the wires in the circuit would have overheated and might have caused a fire. Putting in the penny would restore the excessive current, but the penny would not burn out and the wires would overheat and possibly cause a fire.
17. No, inserting two fuses would be useless because the circuit will already be broken if one fuse blows.
18. Sample answer: Beaches and golf courses tend to be flat and have few tall objects. Lightning tends to be attracted to the tallest object. The charges from the lightning could jump to your body, providing a path for the charges to move to Earth.
19. If one bulb burns out, there will be an open circuit, therefore the other bulb will also go out.

Chapter Test C

1. C
2. C
3. B
4. A
5. A
6. C
7. B
8. A
9. D
10. C
11. A
12. B
13. electrolyte
14. resistance
15. photocell
16. electrode
17. law of electric charges
18. thermocouple
19. energy source
20. B
21. A

Standardized Test Preparation

READING

Passage 1

1. B
2. F
3. C

Passage 2

1. D
2. I

INTERPRETING GRAPHICS

1. A
2. H
3. C
4. F

MATH

1. C
2. I
3. D

Vocabulary Activity

Across

3. parallel
4. insulator
5. current
6. potential
7. charge
11. load
15. thermocouple
17. battery
18. circuit
19. resistance
20. switch

Down

1. cell
2. discharge
3. photocell
5. conduction
8. induction
9. force
10. power
12. conductor
13. static
14. voltage
16. series

SciLinks Activity

1. Sample answer: A diode is a semiconductor that allows electricity to flow in one direction but not the other like a one way valve for electricity.
2. Sample answer: A crystal radio set is able to detect radio signals without a power supply. It includes an antenna and a very long grounded wire that picks up the waves and passes them through the set as an electric current. The current passing through the crystal causes a solenoid to move, which in turn moves a metal plate. The vibrations of this plate produces faint sound waves.
3. Sample answer: Stick a metal plate made of zinc into the potato or lemon. Place a copper plate on the opposite side. Connect a small light bulb or light emitting diode to the two electrodes with copper wire to complete the circuit and observe what happens.
4. Sample answer: A compass consists of a small lightweight magnetic needle balanced on a pivot point. It is aligned with Earth's magnetic field. If you wrap a wire around a compass and place it in a complete circuit, the direction of the compass needle will change.

Lesson Plan

Chapter Opener

Pacing

Regular Schedule: **with lab(s):** N/A   **without lab(s):** 1 day

Block Schedule: **with lab(s):** 1 day   **without lab(s):** N/A

KEY

SE = Student Edition   **TE** = Teacher's Edition

CRF = Chapter Resource File

- **Chapter Starter Transparency** Use this transparency to introduce the chapter.

- **Start-up Activity, Stick Together, SE** This activity helps students see how a pair of electrically charged objects interact.

- **Pre-Reading Activity, Layered Book, SE** Have students create the FoldNote, Layered Book, before reading the chapter.

- **Parent Letter** This letter describes the content of the chapter and encourages parental participation.

Lesson Plan

Section: Electric Charge and Static Electricity

Pacing

Regular Schedule: **with lab(s):** N/A   **without lab(s):** 1 day

Block Schedule: **with lab(s):** N/A   **without lab(s):** 0.5 day

Objectives

1. Describe how charged objects interact by using the law of electric charges.

2. Describe three ways in which an object can become charged.

3. Compare conductors with insulators.

4. Give two examples of static electricity and electric discharge.

KEY

SE = Student Edition   **TE** = Teacher's Edition

CRF = Chapter Resource File

FOCUS *(5 minutes)*

_ **Bellringer, TE** Have students write a definition of the term electric charge.

_ **Bellringer Transparency** Use this transparency as students enter the classroom and find their seats.

MOTIVATE *(10 minutes)*

_ **Activity, Balloons and Static Electricity, TE** Distribute two inflated balloons and a piece of wool cloth to pairs of students. Instruct students to create static electricity by rubbing their balloons against the cloth. **(GENERAL)**

TEACH *(20 minutes)*

_ **Reading Strategy, Reading Organizer, SE** Have students outline the chapter as they read. **(GENERAL)**

_ **Directed Reading A, CRF** These worksheets reinforce basic concepts and vocabulary presented in the lesson. **(BASIC)**

_ **Directed Reading B, CRF** These worksheets reinforce basic concepts and vocabulary presented in the lesson. **(SPECIAL NEEDS)**

_ **Activity, Exploring Charge, TE** Have students tie two balloons to strings and rub them with a cloth to see how the balloons interact with one another. **(BASIC)**

_ **Reading Strategy, Prediction Guide, TE** Students attempt to explain the terms friction, conduction, and induction before reading passages about each of these terms. **(BASIC)**

_ **Demonstration, Charging Objects, TE** As your read, use objects to demonstrate the three ways an object can become charged. **(GENERAL)**

_ **Activity, Electron Transfer, TE** Use chalk, an eraser, and the chalkboard to demonstrate the "electron trail" of chalk left behind when you erase. **(GENERAL)**

_ **Discussion, Charge and Insulation, TE** Lead students in a discussion about the role of insulation in maintaining an object in a charged state. **(GENERAL)**

_ **Connection to Environmental Science, Painting Cars, TE** Scientists research how charge and electric force are used by carmakers to paint cars and how this process protects the environment. **(GENERAL)**

_ **Research, Types of Lightning, TE** Students research the many forms of lightning. **(ADVANCED)**

_ **Going Further, Franklin's Kite, TE** Students research Benjamin Franklin's kite-flying experiment. **(GENERAL)**

_ **Connection to Social Studies, Benjamin Franklin, SE** Students make a poster describing the procedures used by, terms coined by, and inventions related to electricity that were designed by Benjamin Franklin. **(GENERAL)**

_ **Teaching Transparency, Structure of an Atom** Use this transparency to review the structure of an atom.

_ **Teaching Transparency, Law of Electric Charges** Use this transparency to review the law of electric charges.

_ **Teaching Transparency, How Lightning Forms** Use this transparency to review the basics of lightning formation.

_ **Quick Lab, Detecting Charge, SE** Students make a simple electroscope to detect charges in objects. **(GENERAL)**

_ **Datasheet for Quick Lab, CRF** Students use the datasheet to complete the Quick Lab. **(GENERAL)**

_ **LabBook Lab, Stop the Static Electricity!, SE** Students learn how static electricity builds up on objects. **(GENERAL)**

_ **Datasheet for LabBook Lab, CRF** Students use the datasheet to complete the LabBook lab. **(GENERAL)**

_ **Interactive Explorations CD-ROM, Tunnel Vision** Students advise a young inventor about the best circuit to use on his bicycle helmet. **(GENERAL)**

_ **Interactive Explorations CD-ROM Worksheet, Tunnel Vision** Students us the worksheet to complete the exercise. **(GENERAL)**

_ **Reinforcement, Charge!, CRF** Use this worksheet to reinforce basic concepts and vocabulary presented in the lesson. **(BASIC)**

_ **Whiz-Bang Demonstrations, Hoop It Up** Students observe how the repulsion of like charges causes a Mylar™ hoop to hover over an aluminum pan. **(ADVANCED)**

_ **Whiz-Bang Demonstrations, Bending Water** Students learn how static electricity can bend a stream of water. **(ADVANCED)**

_ **Vocabulary and Section Summary, CRF** Students write definitions of key terms and read a summary of section content. **(GENERAL)**

CLOSE *(10 minutes)*

_ **Section Review, CRF** Students answer end of section vocabulary, key ideas, critical thinking, and interpreting graphics questions. **(GENERAL)**

_ **Reteaching, Figuring Out the Charge, TE** Have students compare the charge of an object to their feelings. **(BASIC)**

_ **Quiz, TE** Students answer 2 questions about electric charge and static electricity. **(GENERAL)**

_ **Section Quiz, CRF** Students answer 7 objective questions about electric charge and static electricity. **(GENERAL)**

_ **Alternative Assessment, Concept Mapping, TE** Have students make a concept map that identifies and briefly explains the three methods of charging. **(GENERAL)**

_ **Homework, Static Cling, TE** Have students observe the effect of static electricity in the clothes dryer at home. **(GENERAL)**

Lesson Plan

Section: Electric Current and Electrical Energy

Pacing

Regular Schedule:	**with lab(s):** N/A	**without lab(s):** 1 day
Block Schedule:	**with lab(s):** N/A	**without lab(s):** 0.5 day

Objectives

1. Describe electric current.

2. Describe voltage and its relationship to electric current.

3. Describe resistance and its relationship to electric current.

4. Explain how a cell generates electrical energy.

5. Describe how thermocouples and photocells generate electrical energy.

KEY

SE = Student Edition   **TE** = Teacher's Edition

CRF = Chapter Resource File

FOCUS *(5 minutes)*

_ **Bellringer, TE** Have students answer the question "What is the difference between something that is direct and something that is alternating?"

_ **Bellringer Transparency** Use this transparency as students enter the classroom and find their seats.

MOTIVATE *(10 minutes)*

_ **Discussion, TE** Show students a photo or illustration of the inside of a car battery or D cell and ask them how they think a battery generates electrical energy. (**GENERAL**)

TEACH *(20 minutes)*

_ **Reading Strategy, Reading Organizer, SE** Have students make a table comparing electric currents, voltage, and resistance. (**GENERAL**)

_ **Directed Reading A, CRF** These worksheets reinforce basic concepts and vocabulary presented in the lesson. (**BASIC**)

_ **Directed Reading B, CRF** These worksheets reinforce basic concepts and vocabulary presented in the lesson. (**SPECIAL NEEDS**)

_ **Inclusion Strategies, TE** Students role-play the differences between AC and DC.

_ **Connection to History, Count Volta, TE** Students learn about Count Volta and the development of the battery. (**GENERAL**)

Lesson Plan *continued*

_ **Connection to Biology, Help for a Heart, SE** Students learn about the importance of pacemaker cells in the human heart. (**GENERAL**)

_ **Research, Alternative Energy, TE** Encourage students to research and write about one alternative way of producing electrical energy, such as using wind or water power. (**ADVANCED**)

_ **Discussion, Electrician, TE** Invite a local electrician or electrical engineer to visit the classroom to talk about working with and around electric current. (**GENERAL**)

_ **Reading Strategy, Prediction Guide, TE** Students answer 3 questions before reading the section on resistance. (**GENERAL**)

_ **Group Activity, Modeling Resistance, TE** Instruct each group to create a model or a poster that illustrates the factors that affect resistance. (**GENERAL**)

_ **Activity, Lemon Cells, TE** Have students construct a lemon cell. (**ADVANCED**)

_ **Connection to Life Science, Nerves, TE** Use the "What's in a Nerve?" teaching transparency to teach students more about nerves and electrolytes. (**GENERAL**)

_ **Teaching Transparency, How a Cell Produces an Electric Current** Use this transparency to review how a cell produces an electric current.

_ **Teaching Transparency, Link to Life Science, What's in a Nerve?**

_ **LabBook Lab, Potato Power, SE** Students create batteries using potatoes. (**BASIC**)

_ **Datasheet for LabBook Lab, CRF** Students use the datasheet to complete the LabBook lab.

_ **Calculator-Based Labs, Lemon "Juice"** Students create a battery using a lemon as a cell and four different electrodes. (**ADVANCED**)

_ **Vocabulary and Section Summary, CRF** Students write definitions of key terms and read a summary of section content. (**GENERAL**)

CLOSE *(10 minutes)*

_ **Section Review, CRF** Students answer end of section vocabulary, key ideas, critical thinking, and interpreting graphics questions (**GENERAL**)

_ **Reteaching, Resistance, TE** Use Ohm's law to help students understand resistance. (**BASIC**)

_ **Quiz, TE** Students answer 2 questions about electrical current and electrical energy. (**GENERAL**)

_ **Section Quiz, CRF** Students answer 10 objective questions about electrical current and electrical energy. (**GENERAL**)

_ **Alternative Assessment, Making Models, TE** Supply students with materials to construct a model of a cell. (**GENERAL**)

Lesson Plan

Section: Electrical Calculations

Pacing

Regular Schedule:	**with lab(s):** N/A	**without lab(s):** 1 day
Block Schedule:	**with lab(s):** N/A	**without lab(s):** 0.5 day

Objectives

1. Use Ohm's law to calculate voltage, current, and resistance.

2. Calculate electric power.

3. Determine the electrical energy used by a device.

4. Compare the power ratings of different electrical appliances.

Indiana Academic Standards for Science

8.2.2 Determine in what units, such as seconds, meters, grams, etc., an answer should be expressed based on the units of the inputs to the calculation.

8.2.7 Participate in group discussions on scientific topics by restating or summarizing accurately what others have said, asking for clarification or elaboration, and expressing alternative positions.

8.3.20 Compare the differences in power consumption in different electrical devices.

KEY

SE = Student Edition   **TE** = Teacher's Edition
CRF = Chapter Resource File

FOCUS (*5 minutes*)

_ **Bellringer, TE** Ask students to respond to the question "How fast is a nanosecond?"

_ **Bellringer Transparency** Use this transparency as students enter the classroom and find their seats.

MOTIVATE (*10 minutes*)

_ **Demonstration, Current and Amps, TE** Set up a 100 W light bulb at the front of the classroom and explain to students that the amount of current in the bulb is 1 amp. Ask students to guess how many amps other objects use. (**GENERAL**)

TEACH (*20 minutes*)

_ **Reading Strategy, Paired Summarizing, SE** Have students read the section the section in pairs and then summarize and discuss the material. (**GENERAL**)

_ **Directed Reading A, CRF** These worksheets reinforce basic concepts and vocabulary presented in the lesson. (**BASIC**)

_ **Directed Reading B, CRF** These worksheets reinforce basic concepts and vocabulary presented in the lesson. (**SPECIAL NEEDS**)

_ **Connection Activity, Math, Electric Power, TE** Have students use the equation for power to solve problems. (**GENERAL**)

_ **Math Focus, Using Ohm's Law, SE** Students solve 3 problems using Ohm's Law. (**GENERAL**)

_ **Math Focus, Power and Energy, SE** Students solve 3 problems using the equation for power and the equation for electrical energy. (**GENERAL**)

_ **School to Home Activity, Saving Energy, SE** Students describe 2 ways to save energy with lights in their home. (**GENERAL**)

_ **Critical Thinking, Potentially Shocking, CRF** Use this worksheet to build critical thinking skills. (**ADVANCED**)

_ **Math Skills for Science, Multiplying Whole Numbers** Use this worksheet to help build and reinforce student knowledge about multiplying whole numbers. (**GENERAL**)

_ **Vocabulary and Section Summary, CRF** Students write definitions of key terms and read a summary of section content. (**GENERAL**)

_ **Activity, Power Ratings, TE** Students determine electrical energy usage during a week. (**GENERAL**)

_ **Group Activity, Energy Savers, TE** Have students brainstorm ways to use less energy. (**ADVANCED**)

_ **Connection Activity, Math, Converting Watts, TE** Have students practice conversion of power ratings of common light bulbs. (**GENERAL**)

_ **Connection to Real World, Choose Your Energy, TE** Explain to students that some electric give users the option to choose alternative energy sources. (**GENERAL**)

_ **Activity, Write a Story, TE** Have students write a story about a family losing electricity to their home. (**ADVANCED**)

_ **Connection to Real World, Disposing of Fluorescent Bulbs, TE** Explain to students that fluorescent light bulbs must be disposed of safely. (**GENERAL**)

CLOSE (*10 minutes*)

_ **Section Review, CRF** Students answer end of section vocabulary, key ideas, critical thinking, and interpreting graphics questions (**GENERAL**)

_ **Reteaching, Converting Watts, TE** Have students practice converting the power ratings of common light bulbs. (**BASIC**)

_ **Quiz, TE** Students answer 2 questions about mirrors and lenses. (**GENERAL**)

_ **Section Quiz, CRF** Students answer 5 objective questions about electrical calculations. (**GENERAL**)

Lesson Plan *continued*

_ **Alternative Assessment, Electrical Energy, TE** Have students write a paragraph about electrical energy. **(GENERAL)**

_ **Reteaching, TE** Have students create a concept map. **(BASIC)**

_ **Activity, Replacing Bulbs, TE** Ask students to calculate the electrical energy used in their room each year. **(ADVANCED)**

Lesson Plan

Section: Electric Circuits

Pacing

Regular Schedule: **with lab(s):** 2 days   **without lab(s):** 1 day

Block Schedule: **with lab(s):** 1 day   **without lab(s):** 0.5 day

Objectives

1. Name the three essential parts of a circuit.

2. Compare series circuits with parallel circuits.

3. Explain how fuses and circuit breakers protect your home against short circuits and circuit overloads.

Indiana Academic Standards for Science

8.3.19 Investigate and compare series and parallel circuits.

> **KEY**
> **SE** = Student Edition   **TE** = Teacher's Edition
> **CRF** = Chapter Resource File

FOCUS (5 minutes)

_ **Bellringer, TE** Have students discuss the answer to the question "What happens when you turn the lights on?"

_ **Bellringer Transparency** Use this transparency as students enter the classroom and find their seats.

MOTIVATE (10 minutes)

_ **Demonstration, Circuit Building, TE** Create a circuit, and demonstrate how it works to the class. **(GENERAL)**

TEACH (65 minutes)

_ **Reading Strategy, Brainstorming, SE** Have students brainstorm words and phrases related to electric circuits.

_ **Directed Reading A, CRF** These worksheets reinforce basic concepts and vocabulary presented in the lesson. **(BASIC)**

_ **Directed Reading B, CRF** These worksheets reinforce basic concepts and vocabulary presented in the lesson. **(SPECIAL NEEDS)**

_ **Connection Activity, Language Arts, A Circuit Story, TE** Have students write a short story that includes a circuit in some meaningful way. **(GENERAL)**

_ **Connection to Biology, Nervous Impulses, SE** Students write a one-page paper comparing their nervous system with an electrical circuit. **(GENERAL)**

Lesson Plan *continued*

_ **Inclusion Strategies, TE** Students examine the idea of the flow and interruption of electric current.

_ **Group Activity, Concept Mapping, TE** Students create a concept map to describe what they have learned about electricity. **(GENERAL)**

_ **Teaching Transparency, Parts of a Circuit** Use this transparency to review the parts of an electric circuit.

_ **Quick Lab, A Series of Circuits, SE** Students explore the characteristics of a series circuit composed of a battery and bulbs.**(GENERAL)**

_ **Datasheet for Quick Lab, CRF** Students use the datasheet to complete the Quick Lab. **(GENERAL)**

_ **Quick Lab, A Parallel Lab, SE** Students explore the characteristics of a parallel circuit composed of a battery and bulbs.**(GENERAL)**

_ **Datasheet for Quick Lab, CRF** Students use the datasheet to complete the Quick Lab. **(GENERAL)**

_ **Chapter Lab, Circuitry 101, SE** Students build a series circuit and a parallel circuit and use Ohm's law to calculate the resistance of the circuits. **(GENERAL)**

_ **Datasheet for Chapter Lab, CRF** Students use the datasheet to complete the chapter lab. **(GENERAL)**

_ **Lab Video for Chapter Lab** Use this video to help students better understand the chapter lab.

_ **Long-Term Projects & Research Ideas, The Future Is Electric** Students explore electric cars and other subjects related to the chapter. **(ADVANCED)**

_ **Reinforcement, Electric Circuits, CRF** Use this worksheet to reinforce basic concepts and vocabulary presented in the lesson. **(BASIC)**

_ **SciLinks Activity, Electric Circuits, SciLinks Code HSM0471, CRF** Students research Internet resources related to electric circuits. **(GENERAL)**

_ **Vocabulary and Section Summary, CRF** Students write definitions of key terms and read a summary of section content. **(GENERAL)**

CLOSE *(10 minutes)*

_ **Section Review, CRF** Students answer end of section vocabulary, key ideas, critical thinking, and interpreting graphics questions **(GENERAL)**

_ **Reteaching, Flashcards, TE** Have students create flashcards to help them remember the parts of a series or parallel circuit. **(BASIC)**

_ **Quiz, TE** Students answer 3 questions about electric circuits. **(GENERAL)**

_ **Section Quiz, CRF** Students answer 7 objective questions about electric circuits. **(GENERAL)**

_ **Alternative Assessment, Section Summary, TE** Have students write a one-page synopsis of the concepts covered in this chapter. **(GENERAL)**

_ **Homework, Circuit Drawings, TE** Have students draw two pictures of series circuits, one with two bulbs and one with three bulbs. **(GENERAL)**

_ **Homework, Making a Poster, TE** Have students research and make a poster about series and parallel circuits. **(GENERAL)**

Lesson Plan

End of Chapter Review and Assessment

Pacing

Regular Schedule:	**with lab(s):** N/A	**without lab(s):** 2 days
Block Schedule:	**with lab(s):** N/A	**without lab(s):** 1 day

---

KEY

SE = Student Edition   **TE** = Teacher's Edition
CRF = Chapter Resource File

_ **Chapter Review, CRF** Students answer end-of-chapter vocabulary, key ideas, critical thinking, and graphics questions. (**GENERAL**)

_ **Vocabulary Activity, CRF** Students review chapter vocabulary terms by completing a puzzle. (**GENERAL**)

_ **Concept Mapping Transparency** Use this transparency to reinforce concepts in the chapter. (**GENERAL**)

_ **Brain Food Video Quiz** Use this video resource to assess student's progress. (**GENERAL**)

_ **Chapter Test A, CRF** Assign questions from the appropriate test for chapter assessment. (**GENERAL**)

_ **Chapter Test B, CRF** Assign questions from the appropriate test for chapter assessment. (**ADVANCED**)

_ **Chapter Test C, CRF** Assign questions from the appropriate test for chapter assessment. (**SPECIAL NEEDS**)

_ **Standardized Test Preparation, CRF** Students answer reading comprehension, math, and interpreting graphics questions in the format of a standardized test. (**GENERAL**)

_ **Performance-Based Assessment, CRF** Assign this activity for general level assessment of the chapter. (**GENERAL**)

_ **Test Generator, One-Stop Planner** Create a customized homework assignment, quiz, or test using the HRW Test Generator program. (**GENERAL**)

Other Resource Options

_ **Science in Action Activities, SE** Students explore Math, Language Arts, and Social Studies activities related to the Science in Action features. (**GENERAL**)

_ **Current Science Articles and Activities** Check out articles and activities by visiting the HRW Web site go.hrw.com. Just type in the keyword HP5CS17T.

_ **Internet Activity, HT5DELI, SE** Have students use a computer that has Internet access to complete this activity. (**GENERAL**)

_ **Science Puzzlers Twisters and Teasers, Introduction to Electricity** This worksheet uses the vocabulary and concepts for this chapter as elements of entertaining puzzles. (**GENERAL**)

_ **Science Fair Guide** Use this resource to help guide students and parents though all phases of a science project. (**GENERAL**)

Introduction to Electricity

MULTIPLE CHOICE

1. What are the tiny particles that make up matter?
   a. charges
   b. atoms
   c. electricity
   d. forces

   Answer: B   Difficulty: 1   Section: 1   Objective: 1

2. What is the region around a charged object where an electric force is present?
   a. electric force
   b. proton
   c. electric field
   d. electron

   Answer: C   Difficulty: 1   Section: 1   Objective: 1

3. Which part of the atom has a positive charge?
   a. proton
   b. electron
   c. neutron
   d. nucleus

   Answer: A   Difficulty: 1   Section: 1   Objective: 1

4. What happens when electrons move from one object to another by direct contact?
   a. friction
   b. detection
   c. induction
   d. conduction

   Answer: D   Difficulty: 1   Section: 1   Objective: 2

5. What would you use to see if something is charged?
   a. electric reader
   b. thermometer
   c. electroscope
   d. electric force

   Answer: C   Difficulty: 1   Section: 1   Objective: 2

6. If a material does not allow charges to move through it easily, what is it called?
   a. electrical insulator
   b. electrical conductor
   c. static electricity
   d. electric discharge

   Answer: A   Difficulty: 1   Section: 1   Objective: 3

7. What is the loss of static electricity as charges move off an object?
   a. static
   b. electric discharge
   c. friction
   d. induction

   Answer: B   Difficulty: 1   Section: 1   Objective: 4

8. What does the size of a current depend upon?
   a. protons
   b. electrons
   c. voltage
   d. charge

   Answer: C   Difficulty: 1   Section: 2   Objective: 2

9. What do you call materials with a resistance of $0\ \Omega$?
   a. conductors
   b. superconductors
   c. resistors
   d. photocells

   Answer: B   Difficulty: 1   Section: 2   Objective: 3

10. What is the part of the cell where charges enter or exit?
    a. nucleus
    b. electrolyte
    c. electron
    d. electrode

    Answer: D   Difficulty: 1   Section: 2   Objective: 4

11. As resistance goes up, what happens to the current?
    a. Current goes up.
    b. Current goes down.
    c. Current stays the same.
    d. Current disappears.

    Answer: B   Difficulty: 1   Section: 3   Objective: 1

12. How much energy do fluorescent light bulbs use compared to incandescent light bulbs?
    a. Fluorescent bulbs use more energy.
    b. Fluorescent bulbs use less energy.
    c. Fluorescent bulbs use the same energy.
    d. They can't be compared.
    Answer: A   Difficulty: 1   Section: 3   Objective: 3

13. What is the voltage if the current is 4 A and the resistance is 10 Ω?
    a. 4 V   c. 10 V
    b. 40 R   d. 40 V
    Answer: D   Difficulty: 1   Section: 3   Objective: 1

14. How much electrical energy does a 75 W light bulb use if it is on for 4 hours?
    a. 150 kWh   c. 300 kWh
    b. 150 W   d. 300 W
    Answer: C   Difficulty: 1   Section: 3   Objective: 3

15. In the formula $P = V \times I$, what does the I stand for?
    a. current   c. voltage
    b. intensity   d. resistance
    Answer: A   Difficulty: 1   Section: 3   Objective: 2

16. The power rating of an electrical device is the
    a. amount of energy the device uses.
    b. amount of time a device can be used.
    c. number of watts a device uses.
    d. rate at which the device uses energy.
    Answer: D   Difficulty: 1   Section: 3   Objective: 4

17. Circuits need three basic parts, an energy source, wires, and what else?
    a. charge   c. load
    b. force   d. energy
    Answer: C   Difficulty: 1   Section: 4   Objective: 1

18. How many pathways are there for moving charges in a series circuit?
    a. one   c. three
    b. two   d. four
    Answer: A   Difficulty: 1   Section: 4   Objective: 2

19. In a short circuit, as the resistance decreases, what happens to the current?
    a. increases   c. does not change
    b. decreases   d. drops to 0 A
    Answer: A   Difficulty: 1   Section: 4   Objective: 3

20. What are the small particles that make up matter?
    a. charges   c. atoms
    b. electrons   d. protons
    Answer: C   Difficulty: 1   Section: 1   Objective: 1

21. Why are protons and electrons attracted to each other?
    a. because protons have a positive charge and electrons have a negative charge
    b. because protons have a negative charge and electrons have a positive charge
    c. because they both are positively charged
    d. because they both are negatively charged
    Answer: A   Difficulty: 1   Section: 1   Objective: 1

22. Why is it important that the electrons and protons are attracted to each other?
    a. The attraction determines the size of the atom.
    b. The attraction keeps the electrons from flying away from the nucleus.
    c. The attraction determines the size of the nucleus.
    d. The attraction determines the charge of the atom.
    Answer: B   Difficulty: 1   Section: 1   Objective: 3

23. The size of an electric force depends upon which two things?
    a. the amount of each charge and the size of the electric field
    b. the distance between the charges and the size of the electric field
    c. the number of protons and the distance between the charges
    d. the amount of each charge and the distance between the charges
    Answer: D   Difficulty: 1   Section: 1   Objective: 1

24. What method is involved when charges in an uncharged metal object are rearranged
    without direct contact with a charged object?
    a. friction   c. convection
    b. induction   d. conduction
    Answer: B   Difficulty: 1   Section: 1   Objective: 2

25. Which of the following is positively charged?
    a. protons   c. electrons
    b. neutrons   d. atoms
    Answer: A   Difficulty: 1   Section: 1   Objective: 1

26. Which of the following is NOT an insulator?
    a. air   c. glass
    b. wood   d. copper
    Answer: D   Difficulty: 1   Section: 1   Objective: 3

27. What generates electrical energy from chemical energy?
    a. cell   c. circuit
    b. switch   d. current
    Answer: A   Difficulty: 1   Section: 2   Objective: 4

28. What is the voltage if the current is 0.4 A and the resistance is 3 Ω?
    a. 12 V   c. 1.2 V
    b. 12 Ω d. 1.2 Ω
    Answer: C   Difficulty: 1   Section: 3   Objective: 1

29. A video monitor draws 1.5 A at a voltage of 150 V. What is the power rating of the
    monitor?
    a. 220 W   c. 150 W
    b. 225 W   d. 2,250 W
    Answer: B   Difficulty: 2   Section: 3   Objective: 2

30. How much electrical energy does a 75 W light bulb use if it is on for 5 hours?
    a. 375 W   c. 300 W
    b. 375 kWh   d. 300 kWh
    Answer: B   Difficulty: 2   Section: 3   Objective: 3

31. What is the third part of an electric circuit besides the wires and the load?
    a. force   c. current
    b. voltage   d. energy source
    Answer: D   Difficulty: 1   Section: 4   Objective: 1

32. What is a switch that automatically opens if the current is too high?
    a. fuse   c. circuit breaker
    b. conductor   d. insulator
    Answer: C   Difficulty: 1   Section: 4   Objective: 3

33. Which of these would lower the electrical resistance of a wire?
    a. making the wire thinner
    b. increasing the wire's length
    c. lowering the temperature of the wire
    d. using denser material for the wire
    Answer: C   Difficulty: 2   Section: 2   Objective: 3

34. What happens if you rub a glass rod with a piece of silk and the rod becomes positively charged?
    a. Electrons on the rod are destroyed.
    b. The silk becomes negatively charged.
    c. Protons move to the rod.
    d. The glass attracts more protons.
    Answer: B   Difficulty: 2   Section: 1   Objective: 1

35. When you flip the switch on a flashlight, what is immediately set up?
    a. electric current
    b. voltage
    c. electric field
    d. resistance
    Answer: C   Difficulty: 2   Section: 1   Objective: 1

36. What does the amount of energy released per charge depend upon?
    a. voltage
    b. resistance
    c. current
    d. friction
    Answer: A   Difficulty: 1   Section: 2   Objective: 2

37. In a photocell, what gains energy to move between atoms?
    a. neutrons
    b. electrons
    c. protons
    d. atoms
    Answer: B   Difficulty: 2   Section: 2   Objective: 2

38. What is the rate at which charges pass a given point?
    a. electrical energy
    b. ohms
    c. electric current
    d. voltage
    Answer: C   Difficulty: 1   Section: 2   Objective: 1

39. What goes down as resistance goes up?
    a. Power goes down.
    b. Charge goes down.
    c. Current goes down.
    d. Temperature goes down.
    Answer: C   Difficulty: 1   Section: 2   Objective: 3

40. What is the switch called that opens if the current is too high?
    a. circuit breaker
    b. conductor
    c. fuse
    d. insulator
    Answer: A   Difficulty: 1   Section: 4   Objective: 3

41. What is the equation for electrical energy?
    a. $E = P \times t$
    b. $E = P + t$
    c. $E = P \div t$
    d. $E = P - t$
    Answer: B   Difficulty: 1   Section: 3   Objective: 3

42. An electroscope can determine which of the following?
    a. whether or not an object is charged
    b. the material that a charged object is made of
    c. the strength of the charge on an object
    d. how many electrons are involved in the charge
    Answer: A   Difficulty: 1   Section: 1   Objective: 2

43. Where does most of the electrical energy used in your home come from?
    a. large rechargeable batteries
    b. electric power plants
    c. chemical reactions
    d. small generators
    Answer: B   Difficulty: 2   Section: 3   Objective: 2

44. Who was Georg Ohm?
    a. an electrician
    b. a teacher
    c. an inventor
    d. an author
    Answer: B   Difficulty: 1   Section: 3   Objective: 1

45. Light bulbs may be labeled "100 W" or "40 W." This describes
    a. how long they burn.
    b. how they are disposed of.
    c. how fast the light travels.
    d. how bright they glow.
    Answer: D   Difficulty: 1   Section: 3   Objective: 3

46. The closer together charges are
    a. the smaller the electric force.
    b. the greater the electric force.
    c. the faster the charges move.
    d. the stronger the charges become.
    Answer: B   Difficulty: 1   Section: 1   Objective: 1

47. Good conductors have low
    a. voltage.
    b. resistance.
    c. current.
    d. temperature.
    Answer: B   Difficulty: 1   Section: 2   Objective: 3

COMPLETION

48. A __________________ is a device in a circuit that uses electrical energy to do work.
    Answer: load   Difficulty: 1   Section: 4   Objective: 1

49. Metal cords are often covered in plastic and have metal prongs. This is because metal is a good __________________ and plastic is a good __________________.
    Answer: conductor; insulator
                    Difficulty: 2   Section: 1   Objective: 3

50. In a battery, electric current exists between the two electrodes because there is a difference in __________________ between the electrodes.
    Answer: charge   Difficulty: 1   Section: 2   Objective: 1

51. If the potential difference is increased, the current __________________.
    Answer: increases   Difficulty: 2   Section: 2   Objective: 2

52. Another word for potential difference is __________________.
    Answer: voltage   Difficulty: 1   Section: 2   Objective: 2

53. The electrical outlets in the United States usually supply a voltage of __________________ V.
    Answer: 120   Difficulty: 1   Section: 2   Objective: 2

54. Thin wires have __________________ resistance than thick wires.
    Answer: more   Difficulty: 1   Section: 2   Objective: 3

55. Short wires have __________________ resistance than long wires.
    Answer: less   Difficulty: 1   Section: 2   Objective: 3

56. An object's resistance depends upon the object's material, thickness, length, and __________________.
    Answer: temperature Difficulty: 1   Section: 2   Objective: 3

57. Electric circuits always form a __________________.
    Answer: loop   Difficulty: 1   Section: 4   Objective: 1

Use the terms from the following list to complete the sentences below.

electric current	series circuit
electrical insulators	electric power
electrical	voltage
static electricity	

58. Plastic, glass, wood, and air are examples of good ___________________.
    Answer: electrical insulators

Difficulty: 1	Section: 1	Objective: 3

59. Electrons moving in a wire make up ___________________ and provide energy to the things that you use each day.
    Answer: electric current

Difficulty: 1	Section: 2	Objective: 1

60. Burglar alarms are best wired using a ___________________.
    Answer: series circuit

Difficulty: 1	Section: 4	Objective: 2

61. When the voltage is in volts and the current is in amperes, ___________________ is expressed in watts.
    Answer: electric power

Difficulty: 1	Section: 2	Objective: 2

62. When your clothes come out of the dryer stuck together, they are full of ___________________.
    Answer: static electricity

Difficulty: 1	Section: 1	Objective: 4

Use the terms from the following list to complete the sentences below.

photocell	electrolyte
electrode	resistance

63. A mixture of chemicals in a cell is a(n) ___________________.
    Answer: electrode

Difficulty: 1	Section: 2	Objective: 4

64. The opposition to the flow of electric charge is ___________________.
    Answer: resistance

Difficulty: 1	Section: 2	Objective: 3

65. A device that converts light energy into electrical energy is called a(n) ___________________.
    Answer: photocell

Difficulty: 1	Section: 2	Objective: 5

66. The part of the cell through which charges enter and exit is the ___________________.
    Answer: electrode

Difficulty: 1	Section: 2	Objective: 4

Use the terms from the following list to complete the sentences below.

law of electric charges	energy source
thermocouple	production

67. The law that states that like charges repel and opposite charges attract is called the ___________________.
    Answer: law of electric charges

Difficulty: 1	Section: 1	Objective: 1

68. A device that converts thermal energy into electrical energy is called a(n) ___________________.
    Answer: thermocouple

Difficulty: 1	Section: 2	Objective: 4

69. The three basic parts of a circuit are wires, load, and a(n) ___________________.

> Answer: energy source
>
> Difficulty: 1   Section: 4   Objective: 1

SHORT ANSWER

70. What does Ohm's law tell us?

> Answer:
> Ohm's law gives the relationship between current, voltage, and resistance. It is summarized by the formula $I = V/R$; the current (I) equals voltage (V) divided by resistance (R).
>
> Difficulty: 1   Section: 3   Objective: 1

71. Why is a fluorescent light bulb more efficient than an incandescent light bulb?

> Answer:
> Answers will vary. Sample answer: Incandescent light bulbs are inexpensive, but are inefficient because they reduce mostly thermal energy, not light energy. Fluorescent bulbs use only one-quarter to one-third the amount of energy as incandescent bulbs, and can last up to ten times longer.
>
> Difficulty: 1   Section: 3   Objective: 4

72. How does a circuit breaker help to protect against short circuits and circuit overloads?

> Answer:
> The circuit breaker is a switch that automatically open when the current is too high. A strip of metal in the breaker warms up, bends, and opens the switch, which opens the circuit. The charges then stop flowing.
>
> Difficulty: 1   Section: 4   Objective: 3

73. A string of lights wired together in a series has a burned out bulb. Why do all of the lights go out

> Answer:
> The burned out bulb is a break in the circuit, so the circuit is no longer a closed path through which electric charges may flow.
>
> Difficulty: 1   Section: 4   Objective: 2

74. Find the resistance of a circuit that draws a 1.5 A when 3.0 V are applied. Show your work.

> Answer: By Ohm's law, $R = V/I$. The resistance of the circuit is $3.0 \text{ V}/1.5 \text{ A} = 2 \ \Omega$.
>
> Difficulty: 1   Section: 3   Objective: 1

75. How does clothing that has become charged in a dryer lose its charge?

> Answer: The electric charges are transferred to water molecules in the air.
>
> Difficulty: 1   Section: 1   Objective: 4

76. When you shuffle your feet on the carpet on a dry day, you get a shock from the metal objects that you touch. What is the cause of this?

> Answer:
> The shock is caused by a buildup of static electricity, which is discharged when you touch metal.
>
> Difficulty: 1   Section: 1   Objective: 4

77. How do electric companies usually calculate electrical energy?

> Answer: By multiplying the power in kilowatts by the hours.
>
> Difficulty: 1   Section: 3   Objective: 2

78. How does temperature difference affect the current in a thermocouple?

> Answer: The higher the temperature difference, the greater the current is.
>
> Difficulty: 2   Section: 2   Objective: 5

79. How does the size of the current affect the speed of charges moving in wires?
    Answer: The larger the current is the faster the charges move through the wire.
    Difficulty: 2   Section: 3   Objective: 2

80. If a object touches an electroscope, how can you tell if it is charged?
    Answer: The metal leaves repel each other if the item is charged.
    Difficulty: 2   Section: 1   Objective: 2

81. Briefly explain the relationship between charge and force.
    Answer:
        Charge is a physical property. Objects with a positive or negative charge exert a force
        on other charged objects.
    Difficulty: 2   Section: 1   Objective: 1

82. Discuss the difference between an electrical conductor and an electrical insulator. Give an
    example of each.
    Answer:
        Charges move easily in an electrical conductor but have difficulty moving in an
        electrical insulator. Most metals are conductors. Plastic rubber, and glass are
        insulators.
    Difficulty: 1   Section: 1   Objective: 3

83. What are two factors that affect the amount of electric current in a wire?
    Answer: voltage and resistance
    Difficulty: 1   Section: 2   Objective: 2, 3

84. Explain the difference between wet cells and dry cells.
    Answer:
        Wet cells contain a liquid electrolyte such as sulfuric acid. Dry cells, such as flashlight
        cells, contain solid or pastelike electrolytes.
    Difficulty: 1   Section: 2   Objective: 4

85. According to Ohm's law, what happens to the current if the voltage increases and the
    resistance stays constant?
    Answer: the current increases
    Difficulty: 1   Section: 3   Objective: 1

86. If the current in a wire is 4 A, what is the ratio of the voltage applied to the wire to the
    wire's resistance in ohms?
    Answer: 4:1
    Difficulty: 2   Section: 3   Objective: 1

87. How do electric power companies keep track of how much electrical energy a household
    or business uses?
    Answer:
        Usually they use electric meters that record the kilowatt-hours of energy used by a
        household or a business.
    Difficulty: 1   Section: 3   Objective: 3

MATCHING

    a. cell   e. alternating current
    b. electric current   f. thermocouple
    c. photocell   g. resistance
    d. voltage

88. ____ when charges shift from flowing from one direction to another
        Answer: E   Difficulty: 1   Section: 2   Objective: 1

89. _____ the potential difference between two points in a circuit

    Answer: D   Difficulty: 1   Section: 2   Objective: 2

90. _____ the rate at which charges pass through a given point

    Answer: B   Difficulty: 1   Section: 2   Objective: 2

91. _____ converts light energy into electrical energy

    Answer: C   Difficulty: 1   Section: 2   Objective: 5

92. _____ converts thermal energy into electrical energy

    Answer: F   Difficulty: 1   Section: 2   Objective: 5

93. _____ the opposition to the flow of electric charge

    Answer: G   Difficulty: 1   Section: 2   Objective: 3

94. _____ changes chemical or radiant energy into electrical energy

    Answer: A   Difficulty: 1   Section: 2   Objective: 4

    a. switch   c. series circuit
    b. parallel circuit   d. electric circuit

95. _____ a circuit in which loads are connected side by side

    Answer: B   Difficulty: 1   Section: 4   Objective: 2

96. _____ a complete, closed path through which electric charges flow

    Answer: D   Difficulty: 1   Section: 4   Objective: 1

97. _____ a device used to open and close a circuit

    Answer: A   Difficulty: 1   Section: 4   Objective: 1

98. _____ a circuit in which all parts are connected in a single loop

    Answer: C   Difficulty: 1   Section: 4   Objective: 2

    a. resistance   f. electric current
    b. electrical conductor   g. electric force
    c. friction   h. electric field
    d. voltage   i. fluorescent bulbs
    e. electric discharge   j. thermocouple

99. _____ The higher this is, the lower the current.

    Answer: A   Difficulty: 1   Section: 2   Objective: 3

100. _____ This device uses the temperature difference in wires to convert thermal energy into electrical energy.

    Answer: J   Difficulty: 1   Section: 2   Objective: 5

101. _____ As this increases, so does the current.

    Answer: D   Difficulty: 1   Section: 2   Objective: 2

102. _____ This method of charging happens when you rub a balloon on your hair.

    Answer: C   Difficulty: 1   Section: 1   Objective: 2

103. _____ This is a material, like metal, that allows charges to move easily.

    Answer: B   Difficulty: 1   Section: 1   Objective: 3

104. _____ This is the region around a charged object where a force is exerted on other objects.

    Answer: H   Difficulty: 1   Section: 1   Objective: 1

105. _____ These are more efficient than incandescent light bulbs.

    Answer: I   Difficulty: 1   Section: 1   Objective: 2

106. _____ This is expressed in amps.

    Answer: F   Difficulty: 1   Section: 2   Objective: 1

107. _____ This is the force between two charged objects.

    Answer: G   Difficulty: 1   Section: 1   Objective: 1

108. _____ Lightning is an example of this.

    Answer: E   Difficulty: 1   Section: 1   Objective: 4

a. conduction   c. friction
b. induction   d. electric discharge

109. ____ charging by wiping electrons from one object to another object

Answer: C   Difficulty: 1   Section: 1   Objective: 2

110. ____ rearranging charges in an uncharged metal object without direct contact with a charged object

Answer: B   Difficulty: 1   Section: 1   Objective: 2

111. ____ charging when electrons move by direct contact

Answer: A   Difficulty: 1   Section: 1   Objective: 2

112. ____ losing static electricity as charges move off an object

Answer: D   Difficulty: 1   Section: 1   Objective: 4

a. electrical insulator   c. electrical conductor
b. electric power

113. ____ a material in which charges can move freely

Answer: C   Difficulty: 1   Section: 1   Objective: 3

114. ____ a material in which charges cannot move freely

Answer: A   Difficulty: 1   Section: 1   Objective: 1

115. ____ the rate at which electrical energy is converted into forms of other energy

Answer: B   Difficulty: 1   Section: 3   Objective: 2

ESSAY

116. Why would it be dangerous to replace a blown fuse with a copper penny?

Answer:

The metal strip in the fuse melted because too much current was flowing through the circuit. If the fuse had not blow, the wires in the circuit would have overheated and might have caused a fire. Putting in the penny would restore the excessive current, but the penny would not burn out and the wires would overheat and possibly cause a fire.

Difficulty: 3   Section: 4   Objective: 3

117. As an added safety precaution, would it be useful to insert two fuses in a circuit instead of one? Explain your answer.

Answer:

No, inserting two fuses would be useless because the circuit will already be broken if one fuse blows.

Difficulty: 3   Section: 4   Objective: 3

118. Why could standing on a beach or in the open on a golf course make your body act like a lightning rod?

Answer:

Sample answer: Beaches and golf courses tend to be flat and have few tall objects. Lightning tends to be attracted to the tallest object. The charges from the lightning could jump to your body, providing a path for the charges to move to Earth.

Difficulty: 3   Section: 1   Objective: 3

119. Name something that uses electrical energy that would be difficult to live without. Explain.

Answer:

Accept any reasonable answer. Sample answer: Refrigerators would be very hard to live without because keeping food fresh would be difficult.

Difficulty: 2   Section: 2   Objective: 1

120. Name one positive feature and one negative feature of superconductors.
    Answer:
    Very little energy is wasted when electric charges move in a superconductor.
    However, a large amount of energy is needed to cool them down.
    Difficulty: 2   Section: 2   Objective: 3

121. How does a lightning rod protect a building from lightning strikes?
    Answer:
    The rod is the tallest point on the building, so when the lightning strikes the rod the
    electric charges are carried safely to Earth through the rod's wire that is grounded.
    Difficulty: 2   Section: 1   Objective: 4

INTERPRETING GRAPHICS

Use the diagram below to answer the following questions.

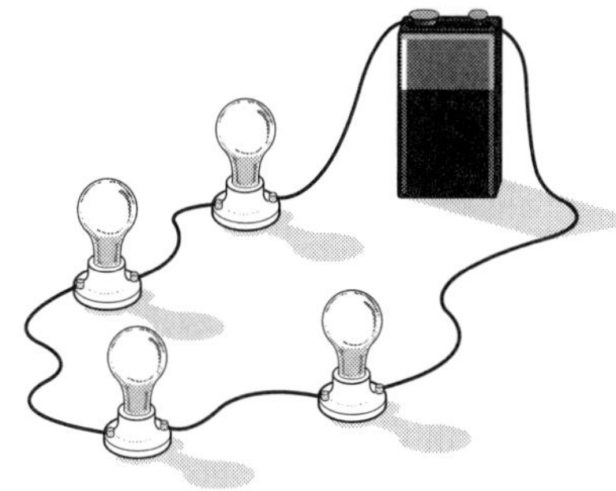

a.

b.

122. _____ parallel circuit
    Answer: B   Difficulty: 1   Section: 4   Objective: 2
123. _____ series circuit
    Answer: A   Difficulty: 1   Section: 4   Objective: 2

Examine the diagram below and answer the question that follows.

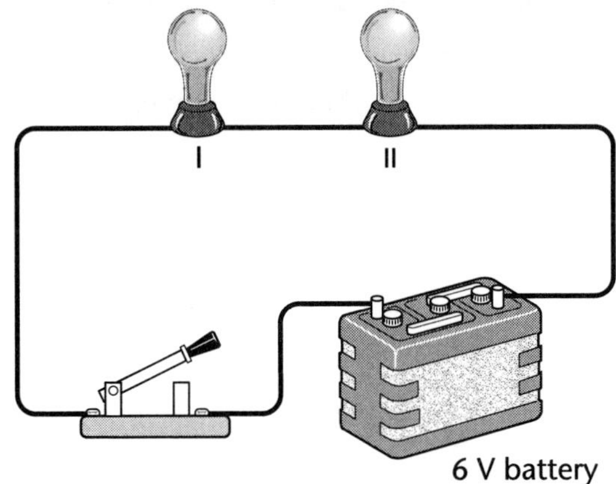

124. If one bulb in this circuit burns out, how will the other bulb be affected? Explain your answer.

Answer:
   If one bulb burns out, there will be an open circuit, therefore the other bulb will also go out.

Difficulty: 2   Section: 4   Objective: 2

TRUE/FALSE

125. ____ If a switch is closed, charges flow freely through the circuit.
   Answer: true   Difficulty: 1   Section: 4   Objective: 1

126. ____ The loads in a parallel circuit do not necessarily all have the same amount of current in them.
   Answer: true   Difficulty: 1   Section: 4   Objective: 2

127. ____ When a short circuit occurs, resistance increases and current decreases.
   Answer: false   Difficulty: 1   Section: 4   Objective: 3